PRACTICAL BUILDING CONSERVATION
VOLUME 3
MORTARS, PLASTERS AND RENDERS

Practical Building Conservation Series:

Volume 1: STONE MASONRY
Volume 2: BRICK, TERRACOTTA AND EARTH
Volume 4: METALS
Volume 5: WOOD, GLASS AND RESINS

PRACTICAL BUILDING CONSERVATION

English Heritage Technical Handbook

VOLUME 3

MORTARS, PLASTERS AND RENDERS

John Ashurst
Nicola Ashurst

Photographs by Nicola Ashurst
Graphics by Iain McCaig

HALSTED PRESS
a division of JOHN WILEY & SONS, Inc.
605 Third Avenue, New York, N.Y. 10158
New York • Toronto

© English Heritage 1988

All rights reserved. No part of this publication may
be reproduced in any form without the prior permission of
John Wiley & Sons, Inc.

Published in the U.S.A. and Canada
by Halsted Press, a division of
John Wiley & Sons, Inc., New York

Library of Congress Cataloging in Publication Data available:
ISBN 0-470-21106-7

Printed in Great Britain

Contents

Foreword — xi
Acknowledgements — xiii

1 Non-hydraulic lime — 1
1.1 *The production of non-hydraulic (high calcium) lime* — 1
1.2 *The lime cycle* — 1
    Slaking
    Slaking lime with sand
    Storing 'coarse stuff'
    The mixing of coarse stuff
    Alternative sources of lime putty
    Hardening of lime mortar and plaster
References — 5

2 Hydraulic limes and cements — 6
2.1 *Hydraulic set* — 6
    PFA
    HTI powder
2.2 *Hydraulic limes and natural cements* — 7
    Hydraulic limes
    Natural ('Roman') cements
2.3 *Artificial cements* — 9
    Portland cement (OPC)
    Other modern artificial cements
        White Portland cement
        Masonry cement
        Sulphate-resisting cement
        High alumina cement (HAC) ('ciment fondu')
        Pozzolanic cements
2.4 *The use of hydraulic limes and cements in building conservation* — 11
References — 11

3	**Mortar additives**	12
3.1	Introduction	12
3.2	Traditional additives	12
	Lime–oil mortar (putty)	
	Lime–tallow mortar	
	Lime–casein	
3.3	Modern additives	13
	Workability aids	
	Air entrainers	
	Accelerators	
	Waterproofers	
	Bonding agents	
4	**External renders**	16
4.1	Types of external renderings	16
4.2	Failures	16
4.3	Identification of faults	17
4.4	Cutting out and matching	18
4.5	Bonding agents	18
4.6	Number of coats	19
4.7	Daywork joints	19
4.8	Repair mixes	19
	For oil mastics, Roman cements, or Portland cement renderings	
	For lime stucco	
	For other original material	
	For very weak material	
4.9	Run mouldings	20
4.10	Repair of daub	20
4.11	Coloured (unpainted) mixes	22
4.12	Painting rendering	22
4.13	Common categories of rendering	26
References		26
5	**Gypsum plasters**	27
5.1	Sources of gypsum	27
5.2	Production, classification and use	27
	Chronology and use	
5.3	Constituent materials	28
	Plaster	
	Hair	
	Other reinforcement material	
	Availability of hair	
	Lath	
	Repairs to lath	
5.4	Decorative plasters	30
	Scagliola	

        Marezzo marble
        Pargetting
        Sgraffito
        Papier-mache
        Carton-pierre
        Gesso
5.5 *Patent cements based on gypsum*   33
        Martin's cement
        Keene's cement
        Parian cement
        Repair
5.6 *Protection of gypsum plaster externally*   34
References   35

6 Plaster ceiling repairs   36
6.1 *The approach to repairs*   36
6.2 *Causes of defects*   37
6.3 *Recording*   37
6.4 *Repairs*   40
        Traditional plaster repair methods
        Developments in plaster repair methods
        Resin repairs
        Treatment of cracks
References   43

7 Limewashes and lime paints   44
7.1 *Use of limewash*   44
7.2 *Constituents*   44
        Limewash constituents and functions
        Lime and lime-tallow wash
        Lime-casein wash
            Ingredients
            Procedure
        Lime–cenosphere (PFA) wash
        Whiting
        Pigments
7.3 *Quantities required*   47
7.4 *Application of limewash*   47
7.5 *Removing limewash*   48
References   48

8 Case study: cleaning and consolidation of the chapel plaster at Cowdray House ruins – phase 1 (1984)   49
8.1 *Background to the project*   49
8.2 *The site*   52
8.3 *Condition of the plaster*   52

8.4	Scope of work in phase 1	52
8.5	Inspection and assessment	53
8.6	General principles and sequence of remedial work	54
8.7	Detailed description of remedial work items	54

    Treatment with biocide
    Consolidation of undercoats with limewater
    Mortar fillets: first stage
    Flush-filling with mortar
    Grouting
        Removal of cement mortar fillets and patches
        Secondary filleting
        Plastering
        Hole filling
    Application of weather coat to undercoat plaster
    Application of mortar fillings and weather coat to finishing plaster
    Second treatment with biocide

8.8	Summary and recommendations	62
8.9	Appendix–Specifications	64

    Mortars
    Grout
    Weather coats
    Biocide
    Flushing solutions
    Procedure for the preparation of brick dust
    Procedure for the preparation of grout

References   66

9 Case study: cleaning and consolidation of the chapel plaster at Cowdray House ruins – phase 2 (1985)   67

9.1	Scope of work in phase 2	67
9.2	Techniques used	68
9.3	Philosophy of approach on modelled figures	69
9.4	South wall window reveal	69
9.5	Linings to altar recess	72
	Cherub's head: altar recess soffit	
9.6	Carbonation experiment	73

10 Case study: cleaning and consolidation of the chapel plaster at Cowdray House ruins – phase 3 (1986)   74

10.1	Background to the 1986 work	74
10.2	Scope of work	74
10.3	Work procedures	75
10.4	Descriptions of the areas of work	75

    The cast frieze (west wall)
    The window head and reveal (north-east wall)
    The panelled area (north wall)

	The large frame (north wall)	
10.5	*Assessment of the Cowdray project*	79

11 Case study: remedial work to secure graffiti on plaster   80
11.1 *Brief*   80
11.2 *Procedure*   80
11.3 *Discussion and recommendations*   81
11.4 *Consolidation*   84
11.5 *Cleaning*   85
      Making good
      Additional notes

FOREWORD

by Peter Rumble CB, Chief Executive, English Heritage

Over many years the staff of the Research, Technical and Advisory Service of English Heritage have built up expertise in the theory and practice of conserving buildings and the materials used in buildings. Their knowledge and advice have been given mainly in respect of individual buildings or particular materials. The time has come to bring that advice together in order to make available practical information on the essential business of conserving buildings – and doing so properly. The advice relates to most materials and techniques used in traditional building construction as well as methods of repairing, preserving and maintaining our historic buildings with a minimum loss of original fabric.

Although the five volumes which are being published are not intended as specifications for remedial work, we hope that they will be used widely by those who write, read or use such specifications. We expect to revise and enlarge upon some of the information in subsequent editions as well as introducing new subjects. Although our concern is with the past, we are keenly aware that building conservation is a modern and advancing science to which we intend, with our colleagues at home and abroad, to continue to contribute.

The Practical Building Conservation Series

The contents of the five volumes reflect the principal requests for information which are made to the Research, Technical and Advisory Services of English Heritage (RTAS) in London.

RTAS does not work in isolation; it has regular contact with colleagues in Europe, the Americas and Australia, primarily through ICOMOS, ICCROM, and APT. Much of the information is of direct interest to building conservation practitioners in these continents as well as their British counterparts.

English Heritage

English Heritage, the Historic Buildings and Monuments Commission for England came into existence on 1st April 1984, set up by the Government but independent of it. Its duties cover the whole of England and relate to ancient monuments, historic buildings, conservation areas, historic gardens and archaeology. The commission consists of a Chairman and up to sixteen other members. Commissioners are appointed by the Secretary of State for the Environment and are chosen for their very wide range of relevant experience and expertise. The Commission is assisted in its works by committees of people with reputation, knowledge and experience in different spheres. Two of the most important committees relate to ancient monuments and to historic buildings respectively. These committees carry on the traditions of the Ancient Monuments Board and the Historic Buildings Council, two bodies whose work has gained them national and international reputations. Other advisory committees assist on matters such as historic gardens, education, interpretation, publication, marketing and trading and provide independent expert advice.

The Commission has a staff of over 1,000, most of whom had been serving in the Department of the Environment. They include archaeologists, architects, artists, conservators, craftsmen, draughtsmen, engineers, historians and scientists.

In short, the Commission is a body of highly skilled and dedicated people who are concerned with protecting and preserving the architectural and archaeological heritage of England, making it better known, more informative and more enjoyable to the public.

ACKNOWLEDGEMENTS

The authors gratefully acknowledge the assistance of Dr Clifford Price, Head of the Ancient Monuments Laboratory, English Heritage, in the reading of the texts.

1 NON-HYDRAULIC LIME

Non-hydraulic lime is the principal binder of most traditional mortars, plasters and renders, although it is widely neglected in modern building practice. In the context of historic building repair an understanding of the material and how to make the most of it is essential. (The hydraulic components of mortars, plasters and renders are considered in Chapter 2, 'Hydraulic limes and cements' and Chapter 3, 'Mortar additives'.

1.1 THE PRODUCTION OF NON-HYDRAULIC (HIGH CALCIUM) LIME

Limestones, including chalk, provide the raw material for lime in Britain. Lime is produced by breaking the stone into lumps and heating the raw material in a kiln. Early kilns were sometimes no more than simple clamps of alternate layers of stone and fuel, covered with clay and ventilated through stoke holes. Traditional kilns, however, are normally flare kilns, in which intermittent burning takes place, or draw kilns, in which loading and burning are continuous. Modern rotary kilns are fuelled by oil or gas, burning the limestone at temperatures between 900°C and 1200°C. The minimum effective temperature for burning limestone for lime is 880°C, but for this temperature to be reached in the centre of the stone lumps, an overall temperature at the surface of 1000°C is necessary. During burning, carbon dioxide (and any water) is driven off. The end product is calcium oxide, 'quicklime', sometimes described as 'unslaked lime' or, rather misleadingly, as 'lump lime'.

1.2 THE LIME CYCLE

The burning of limestone is the first step in the sequence which leads to the setting and carbonation of a lime mortar, render, or plaster. The cycle can be summarized as follows:

Mortars, Plasters and Renders

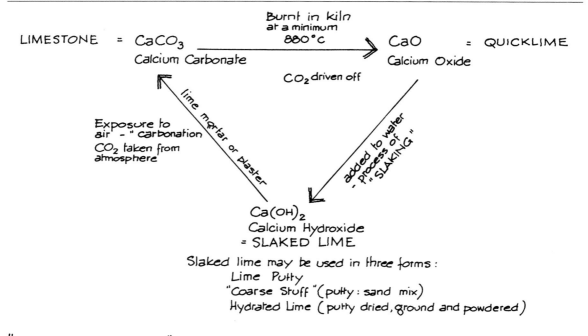

Figure 1.1 The lime cycle

Slaking

Quicklime (unslaked lime) in lump or ground form can be delivered by a number of suppliers while others allow collection of smaller amounts of lump from their works. For site slaking the lime should be delivered as fresh as possible and kept in dry conditions.

Slaking is the reaction of the quicklime with water. If quicklime is left exposed to the air it will absorb water from it and 'air-slake', or 'wind-slake', the calcined lumps gradually reducing to powder with an increase in volume. On site, slaking usually occurs under water.

During the process of slaking, hydroxides of calcium (and magnesium) are formed by the action of water on the oxides. Traditionally, this process was carried out in pits and the slaked lime was left to mature for several months, or even years. Today slaking on site for repair work is most conveniently carried out in a galvanized steel cold water storage cistern. Clean, potable water is run into the tank to a depth of approximately 300 mm and the quicklime is added by shovel (*NB*: the water is *not* added to the quicklime – this is extremely dangerous); because the violent reaction which can occur between water and fresh quicklime frequently raises the water temperature to boiling point, this operation must be carried out slowly and carefully. Eyes must be protected by goggles and hands by suitable gloves, and anyone unprotected in this way must be kept away from the slaking tank. The initial slaking process may be carried out more quickly by first breaking the lumps of quicklime down to a large aggregate size and by using hot water in the tank.

The slaking lime must be hoed and raked and stirred until the visible reaction has ceased; enough water must be used to avoid the coagulation of particles which significantly reduces the plasticity of the lime. Experience will dictate the correct amount of water required, which can be adjusted as the process demands; it is always better to have an excess of water than not enough. The addition of water and quicklime continues until the desired quantity has been slaked. Using an excess of water without 'drowning' the lime results in the formation of a soft, rather greasy mass of material, described as *lime putty*.

Sieving the putty through a 5 mm screen will remove unburnt lumps and the larger coagulations. The screened putty should be left under a few centimetres of slaking water. This 'limewater' may be siphoned off for use in limewater consolidation of friable lime plasters and limestone. (See Volume 1, Chapter 8, 'The cleaning and treatment of limestone by the "lime method"'.)

The lime putty, with a shallow covering of water, should be kept for a minimum period of two weeks before use. Two months is a better period if practicable and there is no upper limit of time. The minimum period is to ensure that the entire mass is thoroughly slaked. After this time, plasticity and workability go on increasing. Old lime putty, which is protected from the air in a pit or bin, acquires a rigidity which is rather like that of gelatin. When the rigid mass is worked through and 'knocked up', it becomes workable and plastic again. This property is peculiar to non-hydraulic lime putty. Any material which has a hydraulic set must not be knocked up after it begins to stiffen.

Slaking lime with sand

A variation on the slaking procedure, which has a long tradition behind it, is to slake the quicklime in a pit, already mixed with the sand which is to be combined as mortar or plaster. The lime needs to be in small lumps so it can be accurately batched by volume against the sand. The process requires time and space and is really only practicable in long programmes of repair or restoration, where it is intended to lay up quantities of lime putty and sand for a long time. The technique has, however, a distinct advantage over more familiar mixing procedures in that this early marriage between binder material and aggregate encourages the covering of all the aggregate particles with a lime paste in a way and to a degree which can never be matched by conventional modern mixing.

Storing 'coarse stuff'

A recommended procedure is to mix the slaked putty with the sand and other aggregates and to store the constituents together, protected from the air as wet 'coarse stuff' for as long as possible to mature. Again, little or no extra water needs to be added as the lime putty contains sufficient to enable it to be worked back to a plastic state. The coarse stuff is the best possible base for mortar and lime plaster, whether or not it is to be gauged later with any pozzolanic additives. Storage (*not* slaking) is best arranged in plastic bins with airtight lids, with an additional covering inside the bin of wet underlay felt, or wet sacks. Another advantage of storing wet coarse stuff is that all the mixing for a large job can be carried out in one or two operations and a consistent mortar or plaster will be available for use as required.

The mixing of coarse stuff

Initial mixing of the coarse stuff and final mixing, or knocking up, must be thorough. But mixing, in the familiar sense of turning over with a shovel, was not considered sufficient in ancient times, nor is it sufficient now, if the best possible performance is to be obtained from the lime mortar. The old practice of chopping, beating and ramming the mortar has largely been forgotten. However, recent field work has confirmed that coarse stuff rammed and beaten with a simply made wooden rammer and paddle or a pick handle, interspersed with chopping with a shovel, significantly improves workability and performance. The value of impact is to increase the overall lime–aggregate contact and to remove surplus water by compaction of the mass.

Alternative sources of lime putty

Some lime suppliers will supply lime putty in plastic sacks by request. If there is no supply available and site slaking is impossible, use hydrated lime and soak it for a minimum period of twenty-four hours in enough clean water to produce a thick cream. For many ordinary building situations, especially if the lime mortar is to be gauged with cement, this practice is quite satisfactory.

Hardening of lime mortar and plaster

When coarse stuff is left exposed to the air, it stiffens and hardens, with a contraction in volume. Lime putty alone undergoes a far greater contraction, and hence is used only to butter very fine joints. Only the minimum amount of water should be added to mature coarse stuff to achieve workability, so that volume changes during drying can be kept to a minimum.

'Carbonation' describes the reaction of the calcium hydroxide with the carbon dioxide in the air, forming calcium carbonate; it is a delicate process dependent on temperature, moisture, the thickness and pore structure of the material with which the mortar or plaster is associated and, of course, the presence of carbon dioxide. There is no chemical 'set' like that of hydraulic limes and cements.

Rapid drying out, which sometimes takes place in hot weather on unprotected work, limits the carbonation process and mortar or plaster in this situation can take little stress and is vulnerable to rain washing. Carbonation should begin while the mortar is still drying out, that is before all drying shrinkage has taken place, and will continue for many years. Soft mortar isolated from the air can remain soft indefinitely.

A chemical indicator is the only sure way of knowing whether carbonation has taken place. In laboratory tests, phenolphthalein can be used for this purpose since it reacts with a sharp red colour to alkaline materials (calcium hydroxide, slaked slime) and is colourless in a neutral or acid environment (calcium carbonate, carbonated lime).

Drying control is normally effected by making sure the joints or wall surfaces are damp before mortar or plaster is applied and by screening the work against strong draughts or heat. In special circumstances carbonation may be significantly accelerated by periodic wetting of the work. This is most conveniently carried out using a hand spray with a fine nozzle which can create a fine mist (not a jet of water). This process is a refinement which has rather a limited application, but it is simple enough to execute for a day or two after the mortar has been placed. Local conditions will dictate the frequency of wetting, but it may be as often as every hour initially if drying out of the face is likely to be rapid, increasing to three or four hours. Water containing carbon dioxide (soda water) has also been used to good effect for this purpose. An experiment use carbon dioxide to advance carbonation is described in section 9.6 of this volume, 'carbonation experiment', p. 73.

REFERENCES

1. Ashurst, John, *Mortars, Plasters and Renders in Conservation*, Ecclesiastical Architects' and Surveyors' Association, London, 1983.
2. Teutonico, Jeanne-Marie, *Architectural Conservation Course, Laboratory Specifications, Tests and Exercises*, ICCROM, Rome, 1984.

See also the Technical Bibliography, Volume 5.

2 Hydraulic limes and cements

2.1 HYDRAULIC SET

Hydraulic mortars harden (set) by chemical reaction with water. No air is needed. Pozzolanic materials produce a hydraulic reaction with slaked lime ($Ca(OH)_2$) due to their reactive silica (SiO_2) and alumina (Al_2O_3). Calcium silicate hydrate, a product of this reaction, forms a network of fibrous crystals or gel which is the main cause of hardening of the mortar (hydraulic set).

This chapter deals with hydraulic lime, to which this setting property is inherent, and with natural cements, artificial cements and other additives which can produce a set when added to non-hydraulic lime. These materials are:

- PFA (pulverized fuel ash)
- Finely powdered brick dust
- HTI powder, prepared from refractory bricks ('HTI'-high temperature insulation)
- Hydraulic lime
- White cement
- Masonry cement
- Ordinary Portland cement

(See also Chapter 3, 'Mortar additives'.)

The phenomenon of 'hydraulic set' seems to have been appreciated first in Mediterranean countries under Roman influence, where there was an abundance of natural materials ejected from volcanoes. These were in the form of rocks such as tuff, trachyte and pumice, or deposits of volcanic ash or earth, such as pozzolana, or trass. Large deposits of ash in the region of Pozzuoli near Naples, used from early times with lime for mortar and later, for Roman concrete and described as 'pozzolana' are still used extensively in Italian and other Mediterranean building industries. Materials similar to pozzolana which produce a hydraulic set are usually described as pozzolanic additives. Roman builders used bricks, tiles and pottery crushed to dust and ground iron slag as pozzolanic additives. All such materials contain reactive silicates which, in the presence of water, react with lime.

Modern practice in Britain makes use of crushed brick dust, HTI powder and PFA of low sulphate content as pozzolanic additives mixed with lime. Yellow brick dust, HTI powder and PFA in the form of light coloured cenospheres (minute glassy bubbles) do not significantly affect the colour of lime mortars, but, of course, red brick dust and grey fuel ash have somewhat limited applications.

While it is still common practice to gauge lime mortar with cement when an initial set is required, other pozzolanic additives are particularly useful where a strong set is not required, as is often the case in the fabric of ancient monuments and historic buildings.

PFA

PFA is a waste product from power stations. Pulverized coal is blown into combustion in a stream of air and burnt. A high percentage of the resultant ash is in the form of minute separate spheres. Seventy-five per cent of this ash is carried away in flue gases and is extracted as pulverized fuel ash ('fly ash' or 'PFA'). Some PFA will react with lime in the presence of water to form a cement-like material (pozzolanic PFA). Mixed with cement, PFA will react with the lime liberated during hydration. Colour, grading and pozzolanicity vary between power stations producing the ash, and even the same station will produce different ash from time to time. Some of the ash is mixed with cement or lime with various other additives for grouting. When ordering PFA it is as well to specify what the ash is to be used for, and to ensure that a low sulphate ash is supplied. A typical sulphate content (as SO_3) of ordinary ash is 1.2 per cent, but for low sulphate ash the amount can be as low as 0.5 per cent and this is preferable by far.

HTI powder

Fireclays are used in the production of ceramic products which are required to withstand high temperatures ('refractories') such as furnace and flue linings. Finely ground material of this kind can be obtained which will react with lime to produce hydraulic properties in mortars. HTI ('high temperature insulation') is most conveniently purchased as a fine powder, rather than a coarse granular material which must be crushed on site.

2.2 HYDRAULIC LIMES AND NATURAL CEMENTS

Hydraulic limes

The source of hydraulic limes is limestone, but limestone which naturally contains a proportion of clay in addition to calcium and magnesium carbonates. Such limestones will yield 'hydraulic' lime after calcination. Other impurities, such as iron and sulphur, may also be present in these limestones. Kilning procedures are the same as for non-hydraulic lime, but the chemical actions are much more complex during the calcination process. As the temperature reaches 900°C, pozzolanic compounds are formed as decomposition of the carbonates and reaction with clay materials proceeds. Over 1000°C calcium aluminates and silicates are formed and sintering will take place, producing a clinker which is somewhat inactive until finely ground. Changes in the firing temperature, as well as in the constituents, can produce hydraulic limes of very different characteristics.

Although many famous hydraulic limes were produced in the UK before the last war and the raw material is still plentiful, no significant quantity of hydraulic lime is now made here. Hydraulic limes are imported from France and are in use on many sites.

The French limes available are pale buff, light grey or white. Not all colours are available at all times. They are delivered in sacks as a finely ground powder. The sacks should be delivered sealed and must be kept dry. The lime must be mixed very thoroughly with the selected aggregates and with the minimum amount of water to make the coarse stuff workable, the mixed material being able to take a 'polish' from the back of a shovel. Mixing should take place on a clean boarded platform before any water is added and then again after watering. This coarse stuff must be used within four hours and must not be knocked up after stiffening has taken place. Correct judgement on the quantity required for each working phase is, therefore, important.

Natural ('Roman') cements

Natural cements are really eminently hydraulic limes. In the eighteenth century various experiments were taking place mixing different limes with volcanic earths. John Smeaton found that Aberthaw (Glamorgan) lime gave better results than others and concluded that the best limes for mortar were those fired from limestones containing a considerable quantity of clayey matter. In the 1790s, the discovery that a useful, quick-setting hydraulic cement could be made by calcining nodules of argillaceous limestone (septarian nodules) resulted in a patent being taken out in 1794 by James Parker of Northfleet. Similar, brown-coloured natural cements were made from the septaria of Harwich and the Solent ('Sheppey' and 'Medina' cements) and Weymouth, Calderwood, Rugby and Whitby. These cements were characterized by their colour and their quick set, which might be as little as half an hour. They were mixed with sand in a 1:1 proportion, sometimes 1:2 and sometimes, for fine moulded work, almost neat. The name 'Roman cement' seems to have been adopted about 1800 and arose from the distinctive pinky-brown colour and hydraulic properties.

Being a strong, durable material it was welcomed as an external rendering and it is in this form that Roman cement is usually found, lined out in imitation of masonry, sometimes coloured with copperas in lime, sometimes painted, rarely left uncoloured. Unfortunately, it was also used extensively for plastic repairs of masonry and for pointing, roles for which it is too impermeable and too strong, and the removal of Roman cement from medieval masonry, especially architectural detail and carving, is one of the most familiar and taxing jobs for the conservator.

A form of Roman cement was available until the 1960s but is no longer made. Repairs seeking to imitate its colour must be of pigmented cement, or lime or, much better, must be based on a plasticized cement relying on carefully selected red and yellow sands to provide the colour.

2.3 ARTIFICIAL CEMENTS

During the production of artificial cements, the essential ingredients of limestone and clay are combined mechanically.

Portland cement (OPC)

In 1811, James Frost took out a patent for an artificial cement obtained by lightly calcining ground chalk and clay together, anticipating the principle which later led to the establishment of many similar 'artificial' hydraulic cements, the most famous of which became known as 'Portland', from its supposed appearance and similarity to the limestone of that name. The beginning of the nineteenth century saw much experiment and investigation into these materials, notably by Vicat in France.

The first Portland cement type in this country was patented by Joseph Aspdin of Leeds, whose plant at Wakefield crushed and calcined a 'hard limestone', mixed the lime with clay and ground the mix into a fine slurry with water. The mixture was fired, broken into lumps and fired a second time. As low temperatures were used, the quality of the cement cannot have been high. By 1838, however, Aspdin's son William was producing the cement at Gateshead and at sites along the River Thames. Brunel used it for his Thames Tunnel in spite of the price being twice that of Roman cement, so that it may be assumed that results were satisfactory and perhaps the calcination was taking place at higher temperatures. To Isaac Johnson belongs the credit, however, of observing that overburnt lumps in the old Aspdin kilns at Gateshead, which he had taken over, made a better final product and were slower setting. At Johnson's works at Rochester, the results of his observations were produced as Johnson's cement. Along the Thames and Medway, a number of cement works opened up, making use of the chalk and the Thames mud and firing at a temperature high enough to produce vitrification.

The cements produced by the late 1850s were close to those produced by modern methods, grinding chalk and clay together in a wet mill and firing the screened slurry at temperatures of 1300° to 1500°C (2372°–2732°F). The chalk is converted into quicklime, which unites chemically with the clay to form a clinker of Portland cement. After regrinding and firing, the white hot clinker is allowed to cool and a small amount of gypsum is added to lengthen the setting time.

Objections to the use of hydraulic limes, natural cements and especially Portland cement are based on their high strength, their rather impermeable character and the risk of transferring soluble salts, especially sodium salts, to vulnerable masonry materials.

Other modern artificial cements

White Portland cement
This cement is produced from chalk and china clay and is burnt using oil fuel instead of coal. The strength of white cement is rather less than the strength of

ordinary Portland cement, but as long as that is recognized, this factor is of no importance in the conservation context and may even be an advantage. White cement is useful in gauging white lime mortars and pale mortar repairs and occasionally in rendering, where the colour of OPC would be wrong. The cement should comply with the requirements of BS 12: 1971; it is about twice the cost of OPC.

Masonry cement
Masonry cements have the advantage over OPC of greater plasticity and greater water retention. They are based on OPC, but have fine inert fillers and plasticizers incorporated. Their principal use is in rendering, where lime might overpower the colour of natural aggregates, or in sandstone repairs, especially 'plastic repairs' where lime has played a role in the decay of the sandstone. The cement should comply with the requirements of BS 5224.

Sulphate-resisting cement
Some situations require the use of a cement which will resist sulphate attack. Some industrial monuments, such as kilns and masonry associated with flue condensates, or sulphate concentration in ground water, are common examples. Sulphate-resisting cement has a reduced tricalcium aluminate content and has good resistance to chemical attack from sulphates. Mixes for mortars and rendering are the same as those based on OPC. Sulphate-resisting cement should comply with the requirements of BS 4027: 1972.

High alumina cement (HAC) ('ciment fondu')
This cement is produced by fusing limestone and bauxite together; it is grey-black in colour and has different properties from OPC. Setting is slow (up to 6 hours for the initial set as against 45 minutes for OPC), workability is good and rapid heat evolution, coupled with early strength development at low temperatures, makes cold weather working less hazardous. Resistance to sulphate attack is good, but to caustic alkalis is poor. The use of antifreeze additives, lime and waterproofers should be avoided. Crushed chalk may be added in place of lime.

High alumina cement is sometimes recommended for repairing Roman or Portland cement stuccos, because, it is claimed, such repairs can be painted at an early stage without the use of special primers, unlike repairs based on OPC, which require special alkali-resistant primers and a long period of waiting to avoid alkali attack on paint. However, it should be realized that the early alkalinity of HAC is not likely to be much less than the alkalinity of OPC (HAC ph 12, OPC ph 12–13) so that special primers are still recommended if early painting is necessary.

The distinctive colour of HAC can be useful in matching black ash mortar. HAC should comply with the requirements of BS 915: 1972. HAC is about three times the cost of OPC.

Pozzolanic cements
Pozzolanic cements in this country are principally mixtures of OPC and pulverized fuel ash. The PFA reacts with lime liberated during the hydration of OPC, to give a slow hardening, low heat cement, with good resistance to sulphates.

2.4 THE USE OF HYDRAULIC LIMES AND CEMENTS IN BUILDING CONSERVATION

The frequent misuse of cements and, less commonly, hydraulic lime should not prejudice against their sensible use in historic building repair and maintenance work. Quite small quantities which should always be specified accurately, will protect lime mortar and rendering against failure during frost. But they are by no means needed in many of the situations where they are habitually employed and should positively be excluded in the vicinity of old lime plaster, wall painting or stone sculpture.

REFERENCES

1 Ashurst, John, *Mortars, Plaster and Renders in Conservation*, Ecclesiastical Architects' and Surveyors' Association, London, 1983.
2 British Standards Institution
   BS 12: 1978 *Specification for Ordinary and Rapid Hardening Portland Cement*
   BS 5224: *Specification for Masonry Cement*
   BS 4027: 1980 *Specification for Sulphate-Resisting Portland Cement*
   BS 915: 1972 (1984) *High Alumina Cement*.

See also the Technical Bibliography, Volume 5.

3 MORTAR ADDITIVES

3.1 INTRODUCTION

See also: This volume, Chapter 1, 'Non-hydraulic lime'; Chapter 2, 'Hydraulic limes and cements'; Volume 2, Chapter 4, 'Pointing stone and brick'.

Lime mortars require time to develop strength and resistance to frost. This limitation to work periods in the north of the northern hemisphere and to work progress generally has inevitably led to trial and error experiments with additives to improve binding and strengthening properties. Setting additives to non-hydraulic lime mortars have been described in the previous chapters. This chapter summarizes other ancient and modern additives which have been used with lime mortar and makes general recommendations.

3.2 TRADITIONAL ADDITIVES

In the British Isles it is likely that additives in ancient times were limited to casein (milk), eggs (whites), linseed oil, fresh blood, beeswax, keratin (from animal hooves and horns), tallow (animal fat), beer, malt and urine. Bitumen was sometimes used as a substitute for or an addition to lime in river and dock works. Waxes, fats and oils introduced some water-repellent properties to mortar; sugary materials reduced the water required and retarded carbonation or set; beer and urine acted additionally as air entrainers.

Most of these additives were in common use for special works after the period of Roman influence and before the experiments with hydraulic mortars in the mid-eighteenth century onwards. Survivors in the nineteenth century from the old practices were largely linseed oil and tallow. Oil and whiting or oil, lime and sand were used as a fine jointing medium in Scotland and the north of England. They are often typified by their white, pasty appearance and fine cross-cracking. Tallow was used as a waterproofing binder with lime and aggregate in marine fortification and dock works; tallow mortars are commonly dense and well-bound even after heavy weathering.

Lime-oil mortar (putty)
The following specification is suggested as a basis for a lime-oil mortar:

- For fine white joints (under 3 mm). 1 part slaked lime, carefully screeded through a fine sieve : 1 part or $\frac{1}{2}$ part fine silver sand. Mix the lime putty and sand together to a stiff consistency. Gauge the paste with boiled linseed oil. The ratio of oil to lime:sand should be approximately 2 litres per m^3 worked to a smooth, stiff paste
- Alternatively, hydrated lime and silver sand may be mixed dry, 1 : 1 or 1 : $\frac{1}{2}$ and gauged direct with enough oil to form the stiff paste
- Traditionally, the surfaces to be filled were primed with linseed oil before placing the mortar

Lime-tallow mortar
The following specification is suggested as a basis for a lime-tallow mortar:

- 1 part fresh quicklime : $2\frac{1}{2}$–3 parts well graded quartz sand and/or flint aggregate. Slake the lime mixed with the aggregates and whilst still hot add shredded or soft tallow (approx. $\frac{1}{4}$ litre to one bucket of quicklime) raking and stirring until all slaking activity has ceased and tallow has dissolved. Turn out, beat, chop and ram the mix, allowing excess water to drain away. A stiff, fatty consistency of mortar is required which will readily hang on the trowel. The mortar should be well packed and ironed into the joints, ensuring good contact with all surfaces. Treatment of the finished joint faces with a flood application of a biocide is recommended to inhibit the formation of mould

Lime-casein
The use of skimmed milk and lime is described in Volume 1, Chapter 8.

3.3 MODERN ADDITIVES

The recreation of the mortars described above is likely to be exceptional. There is rarely any real justification in trying to recreate an ancient mortar by bringing together all the constituents to the extent of trying to trace possible additives and including them again. Even when tracing is possible, it will be costly and will reveal nothing of the methods and techniques used.

Additives to mortar used in the consolidation of historic masonry should be used sparingly and only when there is a specific need. The basic 1:3 lime:aggregate mortar or plaster does not need more than good technique to enable it to perform well for many circumstances, but it will require a setting aid if it is used externally, with less than two months before its first frost. These will be materials such as ordinary brick dust, refractory brick dust ('HTI powder'), pulverized fuel ash, white cement or ordinary grey Portland cement (see Chapter 2, this volume).

Other kinds of additives are only likely to be needed if cement:lime:aggregate mortars or cement:aggregate are used. These are likely to be air-entraining or workability aids.

Workability aids

Lime mortars do not require additives to provide additional workability, but cement:aggregate mixes do. Cement-based mortars are likely to be used in situations where lime is undesirable. Examples may be quoted of exposed corework (especially sandstone) with severe weathering problems and a large surface area of mortar, exposed mortar joints in impermeable granite or basalt masonry, or mortar used in sulphate-rich zones. In some cases the use of a proprietary masonry cement will overcome these problems, but its colour may not be acceptable. White cement and aggregate with an added plasticizer is a possible solution. The plasticizer, or workability aid, is likely to be a 'surface active agent' which reduces the surface tension of the water used in the mix thereby wetting up more easily. Plasticity is similarly improved by entraining small air bubbles. The manufacturers' recommendations on proportions must be carefully followed as these admixtures reduce strength. Other additives such as kaolin and bentonite clays or fly ash act as lubricants and improve cohesiveness, but they must be used sparingly and not to the extent that the water:cement ratio needs to be increased.

Air entrainers

Although some workability aids entrap air, it may be necessary to add a specific air entrainer where resistance to frost is critical. Air entrainment is a characteristic of proprietary masonry cements but where the colour is unacceptable an air-entrained cement:lime:aggregate mix or cement:aggregate mix may be used. Both workability and durability are improved by entraining 4 to 7 per cent of minute, discontinuous and evenly distributed air bubbles. Air-entrained Portland cement is obtainable as an alternative to adding an air-entrainer to the mix. The most common air-entraining agents are alkali salts of wood resins, but they are many and varied.

Standard air entrainment procedures are likely to produce 4 to 7 per cent of entrapped air. The higher percentages have a significant benefit in frost resistance, but it must be remembered that there will be a strength reduction in the mortar and there may be an increase in vulnerability of the stones or bricks if they have only poor or moderate frost resistance themselves. The mortar design must, as always, be determined not just by the exposure risk but by the condition of the masonry generally.

Accelerators

The most commonly used accelerator is calcium chloride (1.5 per cent anhydrous $CaCl_2$ in the gauging water). This must not be used in association with historic masonry because it will be a source of soluble salts. Essential cold weather working must take place within enclosed, heated spaces and be protected by insulating quilts when exposed.

Waterproofers

'Waterproofer' is something of a misnomer applied to additives. Damp-proofing admixtures inhibit water movement by capillary action, while permeability-reducing types seek to prevent water passage under pressure. 'Waterproofed' Portland cement has an addition of calcium stearate or mineral oil. The use of such mortars in historic work is unlikely to be justified and can increase the moisture problems of a masonry wall.

Bonding agents

These are usually PVA (polyvinyl acetate), SBR (styrene butadiene rubber) and acrylic emulsions. They are often made use of to overcome weaknesses in technique on specification and are themselves often not dependable in persistently damp or otherwise aggressive environments. Their use is not recommended except where referred to elsewhere in the text.

4 EXTERNAL RENDERS

4.1 TYPES OF EXTERNAL RENDERINGS

Renderings to walls of historic buildings include a wide range of aggregates, binders and reinforcement. A simple classification of types may be given as follows:

1. Very low-strength daubs, usually applied in a single thick coat to a backing of wattle or lath

2. Low- to medium-strength renderings based on lime applied in two or more coats to a backing of brick, stone, unbaked earth or lath

3. High-strength renderings based on hydraulic cements and applied in two or more coats to brick or stone or lime-based undercoats

4. Medium-strength oil mastics applied in one thin coat to brick, stone, or lime-based undercoats.

All these renderings may be painted. Types (1) and (2) depend for their survival on regular lime washing or painting, although some in type (2) will weather remarkably well without this protection in a sheltered and reasonably dry environment. Types (3) and (4) are frequently painted for aesthetic reasons.

4.2 FAILURES

Most failures of old renderings are attributable to water penetration, usually resulting from lack of maintenance or inadequate protection from the elements. In addition, errors in technique and poor materials obviously play a part. Inappropriate repair methods and materials used to correct failures often contribute to the deterioration they set out to correct. Smooth stucco is the most frequently in need of repair, and is the most difficult to reproduce satisfactorily.

4.3 IDENTIFICATION OF FAULTS

Symptoms	Probable causes
Crazing on surface	• Shrinkage cracking due to dirty aggregate • Excessive early strength • Dense impervious mix • Gypsum added to Portland cement on site (sulphate attack with expansion)
Separation from backing or between coats	• Loss of adhesion due to water penetration • Strong final coat on weaker backing or undercoat • Excessive thickness of coats • Lack of suction control during rendering (inadequate pre-wetting of substrate) • Lack of adequate key
Crumbling and powdering on surface, with or without efflorescence	• Salt contamination from backing, aggregate or rising damp

In some cases the extent of failure may be such that wholescale renewal of the rendering on a wall cannot be avoided. Economics sometimes favour this solution, since it is often cheaper to carry out complete stripping and renewal than to carry out extensive patching. However, since one of the objects of conservation is the retention of the maximum amount of original material and failures are more often local ones, patching techniques are usually desirable on historic work. Important factors when patching are:

- Compatibility of materials
- Matching colour and texture
- Adhesion of the repair

Close examination of the original surface and an overall view of each facade should be made. In the late eighteenth and nineteenth centuries the effects created by the stucco artist were sometimes elaborate and full of subtleties. For instance, the finishing coat was sometimes a composition of different coloured mortars, the colours being changed between the lining out of the artificial joints. The appearance of lime and copperas, or cement and copperas slurries on Roman and Portland cement renders was sometimes enhanced by additional copperas staining to the cement whilst it was still setting. Occasionally the illusion of joints achieved by lining out would be increased by tuck pointing the lines with a ribbon of white lime putty. Unfortunately these decorative effects fall easy victim to weathering or to obliteration by paint. Evidence of colour and unweathered texture should be sought in sheltered zones.

The following recommendations are general, but apply to most repair situations. Pinning and grouting of bulging areas of rendering is described in Chapters 8–10.

4.4 CUTTING OUT AND MATCHING

Areas of rendering which are extensively cracked, or are sounding hollow, should be cut out with sharp chisels to the backing with square edges, or with slightly undercut edges on all but the bottom. When the stucco is lined out in imitation of masonry joints, cutting out should always follow the joint lines, even to the extent of removing some of the soundly adhering material around the cracked or 'live' areas. Stucco which is not lined out should still be cut out to rectangular profiles around the failures, to avoid ragged repairs, working between features of the building if sensibly possible. However carefully this type of patch repair to stucco is carried out the possibility of differences in appearance must be anticipated, but if these are neatly keyed in to the original and matched as closely as possible they will not disfigure the final weathered appearance of the wall in which they are set; if they are ragged or carelessly matched they will always look bad and variations are likely to become more obvious with weathering.

After cutting out, all dust, loosely adherent material, efflorescence and any organic growth must be thoroughly removed by bristle brushing and treatment with a biocide. The area to be re-rendered must be clean, firm and sterile.

Hollows and depressions should be dubbed out in as many coats as necessary, no coat thickness exceeding 10 mm. Before any rendering is applied to a surface, the background must be dampened to reduce and control suction, especially in hot weather. If the substrate is not sufficiently damp it will soak water from the render as it is applied and reduce the effectiveness of the bond and the strength of the render.

Adequate key must be ensured by raking out the joints to 16 mm minimum depth or by hacking the background (rarely necessary unless, as with oil mastic repairs, the background is impregnated with oil) or scoring the preceding undercoat. Where changes in the backing occur, as, for instance, where there are timber frames flush with the infill of brick, or where concrete inserts show on the face, the junction between materials must be bridged with non-ferrous or bitumen-coated galvanized mesh, or stainless steel expanded metal, and covered with a 3 mm–6 mm thick spatter-dash coat. A far less satisfactory solution is to form positive joints and seal on the face with mastic.

4.5 BONDING AGENTS

The usual bonding agents to provide adhesion for rendering to dense backgrounds are polyvinyl acetate (PVA) or styrene butadiene rubber (SBR). Useful as such additives may be they should not be relied on where the wall is likely to remain damp or where excessive salt is present. PVA, in particular, degrades in damp conditions. A mechanical key is always to be preferred to the use of bonding agents used as priming coats, but gauging with SBR may be useful on occasions for bonding to dense backgrounds. Unfortunately, bonding agents are used too readily to cover deficiencies in rendering ability. Where possible it is better to rely on good preparation and sound techniques in application.

4.6 NUMBER OF COATS

Two-coat work is common, but three-coat work is recommended for all but the smallest stucco repairs. The first and strongest coat should be 9 mm – 16 mm thick combed to provide a key for the succeeding coats, each of which should be thinner and of the same strength or weaker than the preceding. Finishing coats should be 6 mm – 10 mm thick. Undercoats should be left at least two days in summer and at least seven days in winter protected by ventilated covers to ensure that the initial shrinkage is over before the next coat is applied. These are minimum times and longer intervals are desirable. Tests for adhesion and strength must be made before a second coat is applied, and the surface lightly sprayed to reduce and control suction immediately before application.

The top coat should be finished with a wood float and/or given a light scraped finish with a fine hacksaw blade and not overworked. Lining out must match exactly the previous pattern, marked out with a rule and marking tool using levels and plumb rules.

4.7 DAYWORK JOINTS

Day-work joints or working lift joints are difficult to disguise, even on painted work. 'Course' lines should be used on lined-out stucco to mask day-work joints, by cutting away some of the previous day's work to 'key in' the new work. Unlined work should, wherever possible, finish every day at some physical break on the building. Where this is not possible, true level lines should be struck off before continuing work, so that no ragged edges show on the finished work. These are important details which must appear in the specification.

4.8 REPAIR MIXES

For oil mastics, Roman cements, or Portland cement renderings
- *Undercoat/s*   1 : 1 : 6   Cement : white lime : sand
              or 1 : 1 : 6   for severe exposure and mouldings
- *Top coat/s*   1 : 2 : 9   Cement : white lime : sand

Mortars, Plasters and Renders

> **For lime stucco**
>
> The range of strengths will vary considerably, and existing standards do not cover the weakest end of the range. For repairing moderately strong lime renders two 1 : 2 : 9 coats as above will be satisfactory
>
> **For other original material**
>
> A final coat and undercoat of hydraulic lime and sand will be more suitable, especially if it is to be limewashed or painted, in the proportions:
> - 2 hydraulic lime : 5 sharp, well graded sand (undercoat)
> - or 1 hydraulic lime : 3 sharp, well graded sand (final coat)
>
> **Ordinary lime : sand stucco**
>
> Should be repaired with the same, that is, 1 : $2\frac{1}{2}$ undercoat (lime putty : sand) and 1 : 3 top coat. Both coats may have a weak pozzolanic additive such as finely powdered brick dust ($\frac{1}{4}-\frac{1}{2}$ part).
>
> **For very weak material**
>
> Where the use of gypsum has been probable a lime gauged with gypsum may be used as follows:
> - 3 part white lime putty : 1 part plaster of Paris*: 6 parts sharp sand
>
> *Gypsum plaster gauged mixes must be limewashed or painted and not mixed with cement.

4.9 RUN MOULDINGS

Smooth stucco finishes tend to be incompatible with recommended standards on suitable sands. The most durable and satisfactory performance is obtained from using a wide range (well graded) aggregate, so that the smaller size grains fill the voids between the larger grains, reducing the percentage of void to a minimum. Coarse material may be unacceptable for the fine stucco, and it will be completely impracticable when mouldings have to be run, as the horsed mould picks up large aggregate and tears and scores the surface as it is moved along the rail. When coarse aggregate has to be omitted, it is very important to take great care in mixing with the lime binder, to ensure that all the aggregate grains are covered with a binder paste. To achieve this, the least possible water necessary to produce a workable mix must never be exceeded.

4.10 REPAIR OF DAUB

Specifications for daub vary enormously in different parts of the country. The analysis, repair and replacement of daub is covered in detail in Volume 2, Chapter 9.

A repair is being run to a length of cornice. The original material was Roman Cement, which can be well matched with a familiar mix such as 1 : 2 : 9, cement : lime : sand. The desired profile is carefully cut out of sheet zinc, which is then firmly mounted in a frame. The plasterer applies the mix to a corbelling out of tile and brick and runs his mould (a 'horsed mould') along horizontal rails and guides fixed above and below the cornice. A plumb line fixed to the mould ensures a true line is maintained along the wall.

4.11 COLOURED (UNPAINTED) MIXES

Colour matching of lime stucco and Roman cement where they are unpainted is one of the most important aspects of their repair. Few limes have any significant colour now, and Roman cement is no longer available. Colour must, therefore, be provided by the aggregates or by staining additives.

Sometime close examination of the original composition will indicate the presence of matchable sands, such as silver sand, which was traditionally used to give a white stone appearance, or special aggregates to give lustre such as crushed glass, spar or granite. Colour additives such as brick dust, earth pigments and, later, red and yellow mineral oxides to produce a stone colour may all be traceable. Much Roman cement stucco was coloured with a fine wash of lime putty, tallow and copperas to simulate Bath stone colour. This coating was usually applied while the Roman cement was damp and therefore stained the outer portion of the render. Because so much colour can be lost and is liable to change, a careful examination of original surfaces should be made under overhangs or in sheltered detail where the most likely survival of original colour and texture will be found.

Every attempt should be made to match surviving original colour with suitable sands, a considerable variety of which are available. Even if an exact match cannot be obtained, the colour of a selected sand may come close enough to the original to be acceptable. Cement:sand mixes with a plasticizer, or masonry cement:sand mixes are often better than cement:lime:sand mixes, because lime tends to lighten the colour of the final stucco too much.

Trial samples should always be set up on the wall against existing surviving material, and judged on their wet and dry appearance.

Particular care should be taken with pigmented mixes to protect the work from rain, and to avoid overworking the finish. In both these situations, separation of the pigment from the lime is a risk, and patchiness and white lime bloom may result. Colours will tend to vary from day to day even with ready-mixed material which can be affected by daily changes in humidity, float type, different pressures on the float and suction and drying conditions. The notes above on patching and day-work joints are especially relevant to pigmented material.

4.12 PAINTING RENDERING

Where colour is to be applied, as long as possible must be allowed to ensure that the new rendering has dried out completely. In average drying conditions, protecting the work from rain and direct sunlight, a 25 mm (1 in) thickness of rendering will take about four weeks to dry. Paint systems which can be applied to new rendering soon after this period are:

- Limewashes
- Lime-casein paints
- Distemper (size-bound)

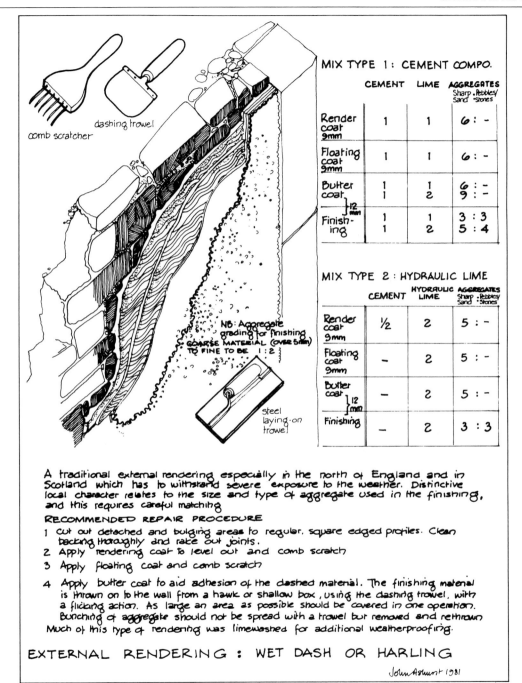

(From 'Mortars, Plasters and Renders in Conservation' Ashurst, J)

Figure 4.1 External rendering: wet dash or harling

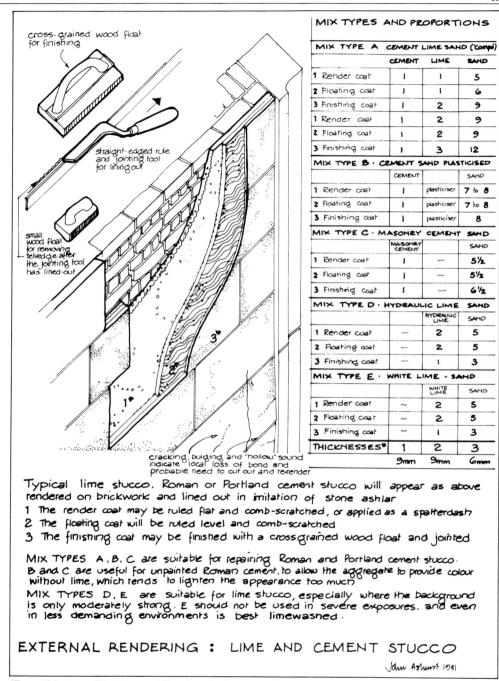

(From 'Mortars, Plasters and Renders in Conservation' Ashurst, J)

Figure 4.2 External rendering: lime and cement stucco

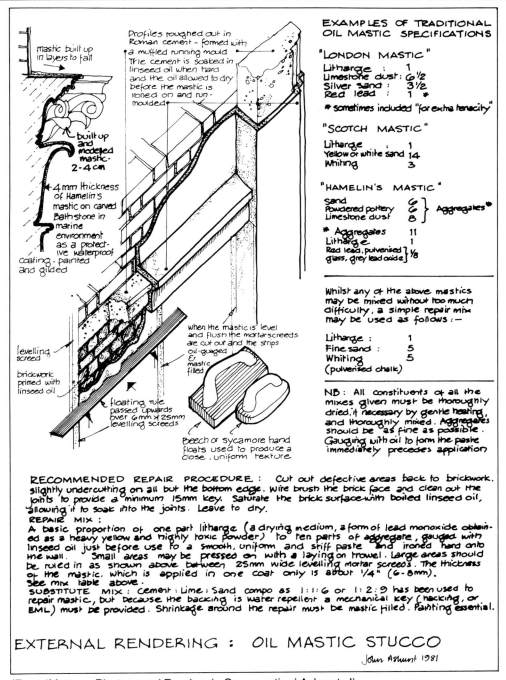

(From 'Mortars, Plasters and Renders in Conservation' Ashurst, J)

Figure 4.3 External rendering: oil mastic stucco

- Distemper (oil-bound)
- Cement paints
- Emulsions
- 'Silicate' paints

These paints are also likely to be the most suitable for matching early surviving examples. Paints which provide a tough impervious envelope especially those which are sprayed on, should always be avoided for historic buildings both on grounds of appearance and because no 'envelope' can ever be complete. Water and salt trapped behind an impervious paint film will result in loss of adhesion and can increase dampness in a building and the risk of persistent deterioration behind the paint.

If any stone is to be painted or limewashed direct, the substrate should be prepared in the same way as for rendering. Similarly the paint system used should be from the list given above. Limewash is likely to be the most usual finish. (See Chapter 7, this volume, 'Limewashes and lime paints'.)

4.13 COMMON CATEGORIES OF RENDERING

Figures 4.1, 4.2 and 4.3 illustrate typical repairs of the common categories of rendering.

REFERENCES

1. Ashurst, John, *Mortars, Plasters and Renders in Conservation*, Ecclesiastical Architects' and Surveyors' Association, London, 1983.
2. British Standards Institution:
   BS 1198, 1199 and 1200: 1976 *Specifications for Building Sands*
   BS 890: 1972 *Building Limes*
   BS 5262: 1976 *External Rendered Finishes*.
3. British Quarrying and Slag Federation Limited, *Lime in Building*, 1968.
4. BRE Digest 196 *External Rendered Finishes*.
5. BRE Digest 197 *Painting Walls: Choice of Paints*.
6. Cement and Concrete Association, *External Rendering*.
7. Department of the Environment *(DOE Advisory Leaflets) (HMSO)*
   1 'Painting New Plaster and Cement'
   6 'Limes for Building'
   15 'Sands for Plaster, Mortars and Rendering'
   27 'Rendering Outside Walls'.

See also the Technical Bibliography, Volume 5.

5 GYPSUM PLASTERS

5.1 SOURCES OF GYPSUM

Gypsum occurs in many different regions often in thick beds between limestone strata and in association with various minerals such as halite (rock salt), calcite and anhydrite. Most gypsum deposits were formed originally by the evaporation of seawater containing a large amount of calcium sulphate in solution. Varieties include a fibrous type which occurs in the form of long parallel silky needles ('satin spar'), or as a fine-grained translucent rock sometimes in massive deposits (alabaster), or as colourless transparent crystals (selenite). Long-exploited sources in this country are the Isle of Purbeck, Dorset, the Knaresborough district, Yorkshire, and the Trent Valley where it was (and to a small extent still is) quarried and mined for alabaster.

5.2 PRODUCTION, CLASSIFICATION AND USE

Gypsum minerals (hydrated calcium sulphate) heated at moderate temperatures yield a hemi-hydrate which is the main component of plaster of Paris.

$$CaSO_4 \cdot 2H_2O \xrightarrow{\text{from } 130°C \text{ in kiln}} CaSO_4 \cdot \tfrac{1}{2}H_2O + 1\tfrac{1}{2}H_2O$$
gypsum → hemi-hydrate + water driven off

Plaster of Paris is obtained by heating the hemi-hydrate at temperatures of between 150° and 160°C

When the calcined material is mixed with water it sets rapidly to form hard crystalline gypsum

$$CaSO_4 \cdot H_2O + 1\tfrac{1}{2}H_2O \xrightarrow{set} CaSO_4 \cdot 2H_2O$$
Plaster of Paris powder + water → hard crystalline gypsum (set plaster)

The set is accompanied by a rise in temperature and expansion of the crystals. Plaster of Paris is the form of gypsum plaster which is found before the developments and patents of the nineteenth century. Gypsum occurs in internal renderings, as gauging in mortars and in floor screeds, and of course in ceiling plaster. Plaster of Paris, used neat, or with lime and/or sand is still used for fibrous plaster work, and for patching and casts (class A plaster).

To make plaster more workable, versatile and durable, a number of innovations were introduced in the nineteenth century. To make plaster more useful as a wall and ceiling covering, retarders such as keratin (obtained from the horns and hooves of animals) were added. Such plasters are described as retarded hemi-hydrate plasters (class B plaster). They are used principally now for sanded undercoats, mixed 1:1–3 with sand.

Higher temperatures will drive off most of the water from the gypsum, and anhydrous calcium sulphate will be formed (anhydrite Class C). Anhydrite occurs naturally, and is burnt and ground with accelerators of set to make anhydrite plaster.

$$CaSO_4 \cdot 2H_2O + \text{Heat } (160°-170°C) \rightarrow CaSO_4 + 2H_2O$$

Such plasters are used neat or with up to 25 per cent lime putty (final coats). Anhydrous gypsum plasters do not react quickly with water, and so an accelerator of set is added, such as alum, potassium sulphate or zinc sulphate. Keene's cement is a material of this kind, and Parian cement is another (class D plaster).

Calcium sulphate is slightly soluble in water, and therefore gypsum-based materials do not weather well in damp climates. Even so, large areas of gypsum plaster have survived considerable periods of exposure where semi-protected, and need careful conservation.

5.3 CONSTITUENT MATERIALS

An appreciation of the materials involved and the approximate periods of their use is necessary not only for placing building activity in its correct historical context, but also to carry out sensible repairs in compatible materials. As much damage has been caused by inexpert or ill advised repair as by the natural agencies of weathering and decay. The following notes on materials should be used as a background guide when considering repairs.

Plaster
Historically speaking, plaster of Paris, 'class A' plaster, with or without lime will be the most accurate material for replacement and repair. However, the modern 'Class A' is undoubtedly a much purer material and a retarded hemi-hydrate plaster gauged with lime is often the best substitute. Techniques of cutting out and replacement of wall plaster are described in Chapter 4, 'External renders'.

Hair
Hair has long been used as a binding material in lime and gypsum plasters. The best hair should be long, strong, and free from grease and other impurities. Ox

Outline Chronology

Prehistoric use The temperature required to burn gypsum to produce mortars and plasters is much lower than that required to produce quicklime by burning limestone. For this reason it was attractive to a country such as ancient Egypt or Greece with an abundance of both limestone and gypsum but with a scarcity of timber for fuel. In Britain gypsum is not found extensively in mortar, although lime mortars are sometimes gauged with gypsum.

c. 1250 Gypsum was introduced into England about this time, imported from massive deposits lying under the Montmatre region. (This importation is associated and probably boosted by the visit of Henry III to Paris in 1254).

Occurs in walls, floors and ceilings in regions where gypsum was plentiful, but most plastering of walls was carried out in lime.

c. 1300 Increasing quarrying and mining of native English deposits of gypsum, especially in the Trent Valley. Red-veined alabaster was probably burnt for plaster, only the white being valued for decorative use.

The plaster was applied to split lath, reed and rush.

From c. 1500 Decorative modelled plaster known as pargetting carried out in gypsum or in lime and gypsum.

From c. 1600 Plastering occurs on imported Red Baltic Fir lath as well as traditional backings.

From c. 1800 Gypsum in common use for internal walls and ceilings, and sometimes for gauging external lime plasters.

Wire netting is introduced in some cases as a backing. Gypsum is used in Marezzo marble.

1834 Improvements on plaster of Paris.
Martin's cement

1838 Keene's cement

1841 Laconte's wire lath for plaster

1846 Parian cement
Sawn lath

1870 Hemp, jute and sawdust as well as hair Selenitic cement

c. 1870 Galvanized wire

From c. 1890 Expanded metal lath.

hair is common, but horse, goat and even human hair have been used as substitutes. The use of human hair is rare because of its fineness and poor strength. Short chopped hair is frequently found, and failures from lumps of hair are sometimes found, where they have caused weak spots. Four kilos (9 lb) of hair to a cubic metre (yard) of plaster is a typical, suitable quantity to provide reinforcement, although up to twice this amount has often been used in the past.

Other reinforcement material
Chopped straw, reed and even grass were commonly used in daubs and sometimes in gypsum-gauged plasters. Reed was used extensively in gypsum plaster floors and in conjunction with lath on stud work and sometimes in ceilings. Hemp and jute were often used in the nineteenth century as alternative reinforcements and sawdust sometimes as an alternative filler.

Availability of hair
Long hair is not easy to obtain for replacement, chiefly because cattle are no longer left to winter outdoors and do not need protective hair. Supplies of ox, horse and goat hair are readily available in Britain. There are restrictions on imported hair from Europe.

Lath
The earliest type of lightweight support for plasters is in the form of interwoven flexible twigs, usually hazel, sometimes ash, woven round willow sticks. Reed bundles were also used horizontally and vertically. From the fifteenth century lath splitting was common, usually cleft oak, sometimes beech and later, very extensively, straight-grained red Baltic fir. From the mid-nineteenth century sawn lath is common with wrought iron nail fixings. Other nineteenth-century innovations are galvanized nails and wire and the use of slates and metal sheet as fireproofing. A wire netting system, stapled to joists, was patented by Edmund Cartwright in 1797. Wire lathing was first used in this country in 1841, and expanded metal lathing from about 1890. Modern lath systems include galvanized and stainless steel expanded metal.

Repairs to lath
Split lath can be obtained in small quantities from a few suppliers. Woven twigs can readily be used following existing patterns, and provide a good key if plastered both sides. The building museums at Avoncroft and Singleton have carried out experiments with wattle, cleft oak, and expanded metal lath, all of which can be seen on site. Sometimes metal lath, bitumen-coated, is an acceptable substitute, or even a modern proprietary profiled sheet lath which also serves as a dry lining.

5.4 DECORATIVE PLASTERS

The repair or restoration of these plasters is an extremely highly skilled operation which must be left to specialist contractors or individuals with experience and expertise.

Scagliola

This is a plaster coloured with pigments and filled with coloured limestone or marble pieces in imitation of true marble breccia or porphyry. It is prepared in strips applied to a background of stone or brick (occasionally wood), polished with a stone and coated with linseed oil. At least three specialist contractors in England are able to reproduce scagliola and a number of conservators are able to repair and conserve.

Marezzo marble

Marezzo is a similar material to scagliola, but relies on pigments and not stone aggregates for effect. Frequently Marezzo was cast on smooth sheets of slate or glass to give a polished surface, and was sometimes used as small areas of external cladding.

Pargetting

Raised and incised decorative external plaster decoration of this type belongs mainly to the sixteenth or seventeenth centuries, or is replacement of designs of that period. Accurate records in the form of photographs, drawings and squeeze moulds should be made before any repair work is contemplated. Considerable craftsmanship is required to undertake modelling work of this kind. Suitable materials are gypsum-gauged lime putty, well-graded sharp sand and long, well-washed hair as reinforcement, sometimes with the addition of a small amount of tallow (clarified animal fat). Linseed oil, wax or, more often, limewash was used to protect the decorated surfaces. Sometimes layers of limewash conceal the sharpness of the original detail and, when partially deteriorated, need to be removed by flaking with a scalpel or the use of an air abrasive pencil; both operations are very specialized and must only be carried out by experienced conservators.

Suitable mixes for repair must obviously take into account the strength and condition of the original, but a 1:1:6 class B plaster:lime-putty:sand is fairly typical, finished with three thin coats of limewash. Victorian stamped work (patterns pressed from blocks) often occurs on buildings of greater antiquity. Typical mixes for this work are strong, for example, 5 parts Portland cement:2 parts lime putty:2 parts sand, and should not be used to repair gypsum or lime pargetted surfaces.

Sgraffito

The art of scratching decoration on plaster, carried out with pointed sticks, is of great antiquity. However, most sgraffito in the UK, especially two- or three-colour work, belongs to the last century. The following specification was used by the Moody students on the old Science and Arts schools in South Kensington, now part of the Victoria and Albert Museum.

The wall was well wetted to control suction on the brick backing, and brought up to levels with the coarse coat. When dry, this in turn was wetted and covered with the colour coat. A thin (maximum $\frac{1}{8}$ in thick) final coat was then trowelled up and the lines of the design cut through to expose the colour coat below.

- Coarse coat : 1 ordinary Portland cement to 3 sharp washed coarse sand
- Colour coat : 1½ ordinary Portland cement to 1 distemper colour
- Final coat (internal) : Parian cement (see following section)
- Final coat (external) : 3 parts Selenitic cement to 2 parts silver sand

The lines were cut with a spatula or trowel or modelling knives to a slanting edge. Part of the skill was in cutting cast shadows to emphasize the design without forming checks to collect water and dirt. The South Kensington example still survives.

Associated with gypsum plasters is the use of papier-mache and carton-pierre. Repair of these materials must again be left to experts.

Papier-mache
From the sixteenth century raised decoration of this kind is found in France and later in England, especially in the eighteenth and nineteenth centuries. Considerable strength may be built up, and thin glazing sections were formed in this way. The principal use was applied moulded ornament used internally. Pressed sandwiches of brown paper and sugar paper were placed together with animal glue or flour pastes and subjected to high pressure in moulds.

Carton-pierre
Carton-pierre is made from pulped paper with glue and whiting (crushed chalk) added. It found considerable decorative use, mainly in the eighteenth and nineteenth centuries, in the form of applied architectural decoration, candelabra and statuary.

Millar (see references) provides the following specification:

- 950 g (2 lb) of animal glue are dissolved in 900 ml (2 quarts) of water to which 250 g (½ lb) of flour and 250 g (½ lb) of paper pulp (dry weight) are added, boiling and beating for one hour
- The mix is then left to simmer for one further hour, when it is poured out onto a slab or mould and stiffened with whiting
- The chalk for whiting should be pure and ground to a fine consistency. 2.7 kg (6 lb) of fine ground chalk are covered with one quart of water, covered and left to stand until it gels. The whiting is then diluted to a creamy consistency

Gesso
Another type of enrichment to plaster may be found in the form of gesso reliefs. A specification for making and applying this (Millar) is as follows:

> - Mix 4 parts of linseed oil with 6 parts melted animal glue or size. Boil the oil and size together. Mix to a creamy consistency whiting with the above. Scratch the plaster surface with the desired design to form a key, and boss out the pattern with cotton wool soaked in gesso
> - A little extra oil is added to finishing coats, and fingers and tools are kept oiled to avoid adhesion. The mix is kept warm whilst working. Finally the work is polished with a wooden spatula or pumice

Specialist plastering contractors are available in the UK to carry out this type of high-class work.

5.5 PATENT CEMENTS BASED ON GYPSUM

The nineteenth century was a period of experiment and innovation, and innumerable specifications exist designed to increase the usefulness of simple plaster of Paris.

The compositions of the most significant three are described here:

> **Martin's cement** (patented 1834)
> Gypsum was soaked in a solution of strongly alkaline pearl ash with a small amount of sulphuric acid added. 450 g (1 lb) of pearl ash was dissolved in 4.5 litres (1 gallon) of water. After thorough soaking, the gypsum was calcined and ground to a fine powder
>
> **Keen(e)'s cement** (patented Greenwood and Keen 1838)
> Plaster of Paris was soaked in a solution of alum. One part of alum was dissolved in one gallon of water heated to 35°C (95°F). Modified Keene's cement is available now as class D plaster
>
> **Parian cement** (patented J Keating 1846)
> Plaster of Paris was soaked in a solution of 1.1 kg (5 lb) borax (sodium borate), 2.2 kg (5 lb) cream of tartar (potassium hydrogen tartrate) in 2.75 litres (12 gallons) of water and subsequently calcined. These cements were high-strength finishing plasters. In particular, Parian was free working and possessed good tensile strength. It was frequently used neat for mouldings over a float coat of 1 part Portland cement to 3 parts sand

Repair
These strong finishes fail principally because the background is too weak for them, and loss of adhesion allows damp penetration. Replastering on a properly keyed substitute background in Keene's cement on strong cement:sand under-

coats will reproduce a similar finish. The use of a key such as a proprietary profiled sheet lath may be necessary. Local failures of mouldings may be cast in plaster of Paris and lime. On major areas of failure the wall is better stripped and a standard plastering specification using class A or class B plasters, with or without lime, substituted.

5.6 PROTECTION OF GYPSUM PLASTER EXTERNALLY

Examples of external gypsum plastering sometimes occur as infill panels in timber-framed buildings in the UK, in areas where alabaster was readily available and easily accessible for extraction. Such plaster may be on lime-ash backgrounds or lime-gypsum backgrounds with a finish of neat gypsum. Unlike external lime rendering the characteristic of these gypsum finishes is often very compact and smooth. The importance of this lies in the vulnerability of gypsum to water. The gypsum must not be allowed to take up water and certainly must not be exposed to running water.

Boiled linseed oil has, in the past, been used to try to protect the finish, but it tends to discolour and attract dirt. An alternative solution is to apply a microcrystalline wax dissolved in acetone or other ketones solvent, applied by brush and finished with a soft cloth. Plaster repair may best be Class A or B plaster mixed with lime putty, 1:1, even if the original was neat gypsum, and the smooth finish is achieved in the traditional way in 3 mm ($\frac{1}{8}$ in) thicknesses in three stages:

1 Application with laying-on trowel
2 Trowelling and floating
3 Compaction by scouring with a cross-grained wood flat lubricated with water

Ornamental plaster is sometimes exposed to external weathering on roofless ancient monuments. This may result in gradual erosion or more drastic failure where water penetration causes loss of adhesion between the plaster and its background.

Every attempt should be made to protect the edges of exposed, surviving plaster. Unfortunately the edges of surviving plaster have often been framed in fillets of strong, cement-gauged lime or neat cement and sand fillets. The relatively impervious nature of these mortars frequently induces a failure by concentrating moisture movements and salt crystallization damage in the critical perimeter zone. These should be substituted wherever possible, by lime mortar fillets. Sometimes a small lead-covered weathering may be used to throw off water above the plaster, but its design must be carefully considered so that water from the weathering cannot be blown onto a lower area of plaster. Pinning of bulging areas may be secured by drilling and fixing with brass screws and washers or lime grouting and pinning. These techniques, including a special grouting technique for plaster is described in Chapter 8, 'Cleaning and consolidation of the chapel plaster at Cowdray House Ruins'.

REFERENCES

1. Ashurst, John, *Mortars, Plasters and Renders in Conservation*, Ecclesiastical Architects' and Surveyors' Association, London, 1983.
2. Building Research Establishment, *BRE Digest 213 Choosing Specifications for Plastering*.
3. British Standards Institution,
   British Standard Code of Practice 211: 1966 Internal Plastering,
   BS 1191: 1973 Part 1 Gypsum Building Plasters.
4. British Quarrying and Slag Federation Limited, *Lime in Building* (Carolyn House, Dingwall Road, Croydon CR0 9XF).
5. Department of the Environment (DOE Advisory Leaflets):
   No 1 Painting New Plaster and Cement
   No 2 Gypsum Plasters
   No 6 Limes for Building
   No 9 Plaster Mixes
   No 15 Sands for Plaster, Mortars and Rendering.
6. Millar, William, *Plastering Plain and Decorative*, Batsford, 1899.

See also the Technical Bibliography, Volume 5.

6 PLASTER CEILING REPAIRS

6.1 THE APPROACH TO REPAIRS

Defects in plaster ceilings may be localized and easily remedied, but crack patterns and distortions may also indicate that the building has problems which have to be identified and solved. The survey of a ceiling with problems may therefore end with an investigation of other parts of a building. Care must always be taken not to treat symptoms without being fully aware of causes. An inspection of an old ceiling will almost always involve lifting boards, and often removing pugging and cleaning out with industrial vacuum cleaners.

Until the 1930s ceilings were commonly of riven or sawn wood laths and three coats of plaster. The first undercoat generally contained ox hair and both undercoats were based on lime and sand or ash, and additionally, in early examples, occasionally contained sugar, sawdust and pounded tiles or other pozzolanic material. Later, the lime was supplemented or replaced by gypsum plaster and, during the last century by Roman or Portland cement. The final coat was of lime putty, gauged with or entirely supplanted by, gypsum plaster.

In repairing existing plasterwork it is extremely unwise to use modern, harder setting types, for when combined with soft old plaster mixes undesirable effects such as shrinkage cracks affecting decorations and condensation, are liable to occur due to variations in the suction and porosity. It is therefore of prime importance that in carrying out repairs to old plaster ceilings and to some extent walls that the material used for patching should match as nearly as possible the nature of the old work.

For patching, mixes based on gypsum plaster, preferably of the retarded hemihydrate type, with the addition of lime are most suitable, as the shrinkage is negligible and the drying time reasonably short.

Most old plaster ceilings show a certain irregularity of surface, so that instead of presenting a perfectly flat appearance in their plain parts, they undulate and may appear to be sagging away from their supports. These irregularities are not necessarily signs of danger.

6.2 CAUSES OF DEFECTS

In the UK, defects in plaster ceilings are generally due to failure of the structure to which the plaster is applied rather than the result of the plaster itself breaking down; but if high levels of condensation or exposure to other forms of damp exist the plaster may become friable and flaky.

Common defects are listed below:

- Sagging may be due to the pulling out or corrosion of nails holding laths, but wood lath and supporting timbers frequently suffer attack by wood-boring beetles or by fungal growths. All sagging and associated cracking must therefore be taken seriously and properly investigated

- Long continuous cracks often occur along the underside of the main supporting beams where the laths have been nailed to the underside of the beam without counter-battens, preventing the formation of squeezed plaster keys

- The ceiling joists, originally tightly fitted, in the course of years shrink and pull out of the mortices, therefore exerting a tensional stress in the plaster beneath with subsequent distinctive cracking

- Floor and roof timbers are also subject to movement and sagging which may cause similar longitudinal cracking with displacement. More seriously, the end of beams buried in external solid walls may have decayed. The retention of valuable ceilings may necessitate in situ repairs of decayed structural timbers (see Volume 4, Chapter 1)

- Discolouration of the plaster or severe flaking of the surface is an obvious sign of dampness from leaking pipes, or roofs, or through walls that are cracked or have defective pointing

- Fine cracks wandering across a ceiling in an irregular way are not necessarily dangerous but when they have widened or when one edge is lower than the other, it is essential to investigate the cause and remedy the defects.

6.3 RECORDING

Photographs are essential before any remedial work is put in hand, supplemented by drawings and sometimes by squeeze moulds. The moulds are necessary for the reproduction of detail casts in plaster, especially where the necessary expertise in modelling matching work is not available. They can be in clay or silicone rubber applied on a barrier of tissue or after the application of a release agent. It is as well to remember that clay will distort as it dries. A plaster cast should be made before this takes place.

Mortars, Plasters and Renders

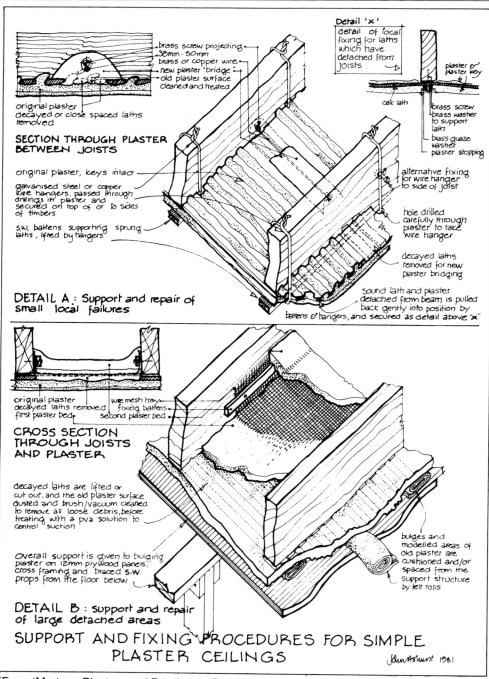

(From 'Mortars, Plasters and Renders in Conservation' Ashurst, J)

Figure 6.1 Support and fixing procedures for simple plaster ceilings

Plaster Ceiling Repairs

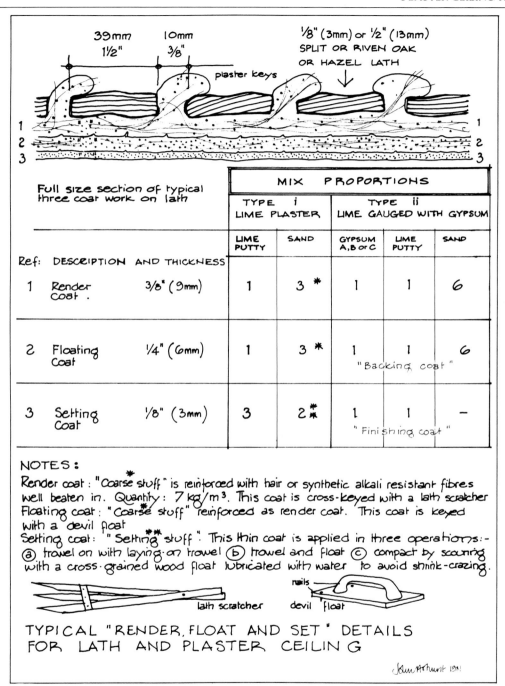

(From 'Mortars, Plasters and Renders in Conservation' Ashurst, J)

Figure 6.2 Typical 'render, float and set' details for lath and plaster ceiling

6.4 REPAIRS

Traditional plaster repair methods

Where the ceilings have sagged due to laths pulling away from the joists or studs, areas can be pulled back and secured with screws using a flat washer to bear directly on the laths, the plaster being cut away for this fixing. A second washer of brass gauze or perforated zinc should also be used to afford a key for plastering in the chase.

Failure through inadequate or ineffectual keying, or lack of adhesion between coats of plaster may be remedied in a number of ways, the suitability of which will be relative to the intrinsic value of the plaster. The traditional procedure of overcoming loss of keys, assuming the sound condition of the structural timber is as follows:

1. Dust and debris is cleared from between the joists using soft brushes and a vacuum cleaner

2. Sagging areas of plaster are supported by plywood panels and bearers and props from the floor below, or by drilling through the plaster at intervals and supporting the panels and bearers with wire loops threaded through the plaster and secured to the joists. In each case the ply panels are separated from the plaster face by pads of underfelt or similar soft materials

3. The condition of the laths is examined to check for rot and inadequate spacing and fixing. Decayed laths or those spaced too close to allow keying are sawn through with a hacksaw blade and snapped off

4. The support for the areas of now unsupported plaster is formed by means of a 'plaster bridge' between the joists where the laths have been removed. The old plaster is cleaned as thoroughly as possible and treated with either a thin coating of size in water, shellac in methylated spirit, or PVA solution (10 per cent) in order to control the suction and prevent rapid removal of water from the new plaster. The solution is applied to the adjacent laths as well as to the plaster. A key is formed on each joist with brass screws left projecting 38–50 mm ($1\frac{1}{2}$ in–2 in) and the area filled with superfine plaster of Paris. The plaster is retarded (mixed with a solution of glue size and water) unretarded plaster of Paris sets in only 6–7 minutes

5. Larger areas of unsupported plaster need more substantial treatment. The back of the plaster is prepared as previously described. Complete support is then provided by forming woven wire 'troughs'. Copper or brass small-mesh wire (6 mm ($\frac{1}{4}$ in) mesh) is prepared in convenient lengths and wide enough to span between the ceiling joists and to turn up at the sides and over the top of the joists. Retarded plaster of Paris is then laid between the joists and worked well into the old plaster. Into this layer

of plaster, the section of wire mesh is gently pressed through and the turned up margins are secured with battens screwed to the joists.

A second coat of plaster is applied to complete the bedding of the new bay before the first layer has set, to make up a total thickness of about 12–18 mm ($\frac{1}{2}$in–$\frac{3}{4}$in) and to completely embed the mesh, forming a splayed fillet against the sides of the joists. If this treatment is carried out over a large area, it is important to allow adequate time for setting and drying out of one section before working on an immediately adjacent section

6   Once the ceiling has been secured in this manner the under surface can be made good and the ornamentation modelled in as necessary. Heavy enrichments should be further secured with heavy gauge copper wire binders or 'hairpins' carried through the background and fastened over the woodwork above

Developments in plaster repair methods

The considerable disadvantage of the plaster-bridge system is the additional weight which has to be carried. This may or may not be a problem, depending on the size and condition of the ceiling framing. The following comparisons give some guidance on weights and other properties of alternative solutions.

	Weight increase %	Drying time	Adhesion
Plaster of Paris + scrim	18	$8\frac{1}{2}$ days	poor/fair
Proprietary internal crack filler + scrim	9.1	8 days	good
Acrylic resin and glass fibre mat	4	15 days	very good
Epoxy and glass fibre mat	4	15 days	very good but non-reversible
(Reference 8)			

Light-weight systems have been developed (8) using acrylic emulsion bonding agents, hydrated lime, glass micro balloons and fluid coke in the proportions:

- 2 parts hydrated lime
- 2 parts glass micro balloons (cenospheres)
- 2 parts 'fluid coke' (also known as 'activated carbon')

- 4 parts acrylic emulsion
- ¼ part water

For this technique the considerable problems of pre-wetting the old, dusty plaster and dry wood are overcome by using a mixture of 3 parts water, 3 parts ethyl alcohol and 2 parts acrylic emulsion bonding agent, applied with a hand pressure spray which produces a jet and subsequent foaming.

All the materials are available in the UK except the 'fluid coke'. This light-weight, black-coloured aggregate has been imported from the USA, but a more readily available light-weight aggregate may be substituted. Alternatively, a light-weight plaster may be extended with cenospheres.

The pre-wetting and light-weight support material are applicable where laths are sound but the plaster has lost its key, in which case the laths must be very carefully drilled (5 mm ($\frac{3}{16}$ in) holes at 300 mm (1 ft) centres) and injected with the materials, or in conjunction with mat or mesh, to substitute for decayed lath. Operatives must wear respirator masks when handling glass micro balloons, and gloves when handling resins.

Resin repairs

An alternative method of treatment where keys are fractured or non-existent is to repair local areas of decayed lath with glass fibre cloth and a two-part resin. After careful vacuum cleaning the joists and the top of the ceiling plaster are treated with a priming coat of two-part resin. Layers of glass fibre cloth are laid over the plaster panels and extended over the nearest sound timber supports. A further impregnation application of the low viscosity resin creates a completely new bond of plaster to timber. This method of support has the advantage of high strength and lightness, and strengthens the timber against any future attack of beetle. Great care is necessary in selecting the right resin system to achieve maximum penetration without risking bleeding of resin through to the underside of the plaster. The disadvantage, which may be considerable, is the permanence and 'irreversibility' of the system. Although the method has been used over quite large ceilings making the whole assembly of joints, lath and plaster homogeneous, there should be good reasons why one of the alternative methods already described cannot be used before this irreversible commitment is made.

Treatment of cracks

Cracks should be filled after the supports have been repaired. If fine they may be filled with a patent crack-filler; if large a lime or vermiculite plaster will be more suitable. The larger cracks should be carefully cut out with a sharp chisel or knife to form a dovetail key for the repair; it is important to avoid any impact which will risk further damage to plaster keys. Before the repair mix is placed, all cracks must be thoroughly cleaned by hand vacuum cleaners or even, if the size of the operation justifies it, with an air abrasive tool and a fine abrasive such as aluminium oxide crystals. After cleaning, the dust should be removed by hand spraying with water or a water and alcohol mixture.

REFERENCES

1. Ashurst, John, *Mortars, Plaster and Renders in Conservation*, Ecclesiastical Architects' and Surveyors' Association, London, 1983.
2. Jack, J F S, *The Repair of Plaster Ceilings*, Internal note, Ministry of Public Buildings and Works, 1946.
3. British Quarrying and Slag Federation Limited, *Lime in Building* (Carolyn House, Dingwall Road, Croydon CR0 9XF).
4. BRE Digest 213, *Choosing Specifications for Plastering*.
5. British Standards Institution:
   British Standard Code of Practice 211: 1966 Internal Plastering
   BS 1198, 1199 and 1200: 1976 Specifications for Building Sands
   BS 1191: 1973 Part 1 Gypsum Building Plasters
   BS 4721: 1971 Ready Mixed Lime: Sand for Mortar.
6. Department of the Environment (DOE Advisory Leaflets):
   No 2 Gypsum Plaster
   No 6 Limes for Building
   No 9 Plaster Mixes
   No 15 Sands for Plaster, Mortars and Rendering.
7. Stagg, W D and Masters, Ronald, *Decorative Plasterwork: Its Repair and Restoration*, Orion Books, 1983.
8. Phillips, M W, *Association for Preservation Technology (APT) Bulletin XII, No. 2*, 1980.

See also the Technical Bibliography, Volume 5.

7 LIMEWASHES AND LIME PAINTS

7.1 USE OF LIMEWASH

Limewash is traditional surface finish for daubs, lime plaster, limestone and earth walls and has also been used on sandstone, brickwork and timber. Multiple applications of limewash may be found on many historic building surfaces, sometimes obscuring detail because of the accumulated thickness. Although once an almost universal treatment, the quality of limewashes varied enormously, some being extremely fine and some thick with a coarse texture. The adhesion of some early limewash to limestone surfaces is quite remarkable and may be evidence of application to surfaces from which quarry sap was still drying. Old limewashed surfaces should be maintained with limewash and suggestions of covering with any alternative paint system strongly discouraged.

7.2 CONSTITUENTS

The basic constituent is lime, to which pigments may be added for colour, and tallow, linseed oil or casein for a more durable treatment. Washes incorporating tallow are normally selected for external treatment, but oil and animal glue were common alternatives. Common salt and alum were sometimes included in traditional recipes. Salt and alum help to emulsify the fat in lime-tallow washes but in remote, marine environments salt water was frequently used as the most readily available water supply. In such situations it would be very pedantic to suggest that new limewash should be salt-free. In almost every other circumstance, however, it is not recommended that salt is included in treatments to porous masonry, because of the risk of damage due to repeated crystallization of soluble salts during wetting–drying cycles. Hydraulic lime may be used for limewashing but has no particular advantage over non-hydraulic and is not available in the UK as quicklime (see reference to 'shelter-coating' in Volume 1, Chapter 8, 'The cleaning and treatment of limestone by the "lime method"').

Limewash constituents and functions

Constituent	Function
Lime	Basic covering medium. Carbonation takes place slowly and 'dusting' may continue for many years, especially if moisture and salt are present. Relatively easily removed by rain
Tallow	Clarified animal fat. Traditional binding and weather-proofing medium
Raw linseed oil	Traditional weather-proofing medium. Little used today
Animal glue	Traditional binding medium. Hardly ever used today
Milk or casein	Traditional additives which react with lime to form relatively insoluble calcium caseinate and give improved resistance to dusting and washing by rain
Common salt / Alum	Added to lime-tallow washes to assist in emulsifying the fat
Trisodium phosphate	Added to lime-casein washes to assist in dissolving the casein. Forms calcium phosphate which is relatively insoluble
Copper sulphate / Formaldehyde	Added to washes containing fats or oils to inhibit the formation of mould
Pigments / 'Washing blue'	Added to colour the white lime or, in the case of the weak blue dye, to impart added brilliance to the white colour

Lime and lime-tallow wash

The simplest wash is made by slaking fresh quicklime, sieving the resulting putty and adding sufficient water to make a thin cream. Ordinary hydrated lime can be used after soaking it for twenty-four hours in water, but its performance without additives is generally accepted to be inferior to washes prepared from freshly slaked lime.

External washes which are gauged with tallow (clarified animal fat) have this ingredient added during the slaking process so that the fat is melted by the heat generated. Raw linseed oil instead of fat may be beaten in with a whisk or electric blender at any time.

Pigments, if required, may be broken up and added at almost any stage, but mixing is facilitated by dissolving the required quantity first in hot water and adding after sieving the putty. Pigments should comply with the requirements of BS 104: 1975 'Pigments for Portland cement and Portland cement products'.

Lime-casein wash

The value of casein depends on its ability to unite with the lime to form the compound calcium caseinate which becomes relatively insoluble on exposure to air. Solution of the casein is aided by the addition of tribasic sodium phosphate.

Ingredients

Non-hydraulic quicklime	5.0 kg slaked in 5 litres of water
or	
Non-hydraulic hydrated lime	12.5 kg soaked in 14 litres of water
Casein	0.9 kg
Trisodium phosphate	0.57 kg
Formalin	0.5 litres
Pigment as required	

Procedure

Soak the casein in hot water for two hours. Ordinary commercial quality casein prepared from separated milk is adequate. Dissolve the trisodium phosphate in two litres of water. Add the pigment to the lime, stirring vigorously. When all solutions are quite cool, add the trisodium phosphate solution to the casein solution and then, as slowly as possible, mix the lime solution in. The formaldehyde must be added, dissolved in seven litres of water, just before use, stirring all the time. Rapid addition of the formalin will result in a gelling of the whole mixture. Thin with water as required. This mix should be used at once and not stored for more than one day.

Lime-cenosphere (PFA) wash

A form of ready-mixed limewash, which has shown itself to be useful over the past seven years, is composed of hydrated lime to which 10 per cent of pozzolanic PFA is added (cenospheres). A polymer binder, intended to take the place of traditional tallow, is also included. Pigment is added at the time of mixing if colour is required.

The proprietary binder is available in bags and only requires to be added to cold water. The limewash is reasonably resistant to rubbing and to normal external exposure. Adhesion is markedly better on porous surfaces such as old lime plaster or brickwork than on fresh gypsum plaster.

Whiting

Whiting is a traditional internal finish prepared by crushing chalk to powder, and mixing it with water and size. It is significantly inferior to limewash and requires frequent maintenance.

Pigments

Lime-fast pigments complying with BS1014 should be used. Trial samples are always advisable. Traditional colours for limewash may be prepared with the following mixes:

> *Cream*: 1.8–2.7 kg (4–6 lb) of ochre to 36.5 litres (8 gallons) lime putty
>
> *Fawn*: 2.7–3.6 kg (6–8 lb) of umber, 0.9 kg (2 lb) Indian red and 0.9 kg (2 lb) lamp black to 36.5 litres (8 gallons) lime putty
>
> *Buff*: 2.7–3.6 kg (6–8 lb) raw umber and 1.35–1.8 kg (3–4 lb) lamp black to 36.5 litres (8 gallons) of lime putty

7.3 QUANTITIES REQUIRED

- Approximately 4.8 litres (1 gallon) limewash will cover 11 m^2 (100 ft^2)
- 4.8 litres (1 gallon) water will slake 4.4 kg (10 lb) of quicklime to which if required, 400 g ($\frac{3}{4}$ lb) of tallow may be added
- Pigment quantities must be established by trial and error on sample patches. The wet colour will always look much stronger than the dry

7.4 APPLICATION OF LIMEWASH

The surfaces to be limewashed should be brushed down to remove loose dust and scale using a stiff, then a soft bristle brush. Any organic growth present such as mould, algae or lichens should be treated with a biocide and brushed or scraped clean. Any making good in the plaster surface should be carried out in lime mortar and fine sand or stone dust. Immediately before application the surfaces should be sprayed with water to avoid too great a suction on the limewash. Limewash which is applied to backgrounds which are too dry will develop fine cracks and become dusty and friable. Application of water is most conveniently carried out with a backpack pneumatic sprayer fitted with a hose and a wand with an adjustable nozzle.

The limewash is applied to the damp surface with a grass brush or other suitable soft-haired brush, working it well into the surface, but not trying to cover cracks and imperfections with the first application. As the limewash dries out it will be semi-transparent. Subsequent applications, each preceded by light wetting of the surface, will build up more body. Three applications should be considered as standard with four or five on very exposed faces. Thick applications will show brush texturing and are liable to craze; thin applications should produce a sound surface with a silky texture.

Limewash should not be allowed to dry out too quickly by exposure to hot sun, artificial heat or strong draughts. Temporary screening should be provided if possible where this risk is likely externally and controllable heat sources restricted. Temporary protection should also be provided to protect fresh limewash from rain.

7.5 REMOVING LIMEWASH

Procedures for the removal of limewash are described in Volume 1, section 5.9.

REFERENCES

1 Ashurst, John, *Mortars, Plasters and Renders in Conservation*, Ecclesiastical Architects' and Surveyors' Association, 1983.
2 Schofield, Jane, *Basic Limewash*, prepared for SPAB, 1984.

See also the Technical Bibliography, Volume 5.

8 Case Study: Cleaning and Consolidation of the Chapel Plaster at Cowdray House Ruins – Phase 1 (1984)

8.1   BACKGROUND TO THE PROJECT

This report describes research and practical field work by the Research and Technical Advisory Service of English Heritage on the decayed decorative plaster of the chapel at Cowdray House ruins, Midhurst, West Sussex, where internal plaster was (and still is) exposed to external weathering conditions.

1984

The project began in 1984 in response to a request to RTAS for a specification for the cleaning, consolidation and maintenance of the chapel plaster. During the first inspection it became clear that there was insufficient knowledge and practical experience for such a specification to be prepared and the work undertaken correctly. It was decided that a three-week exploratory project was necessary to try out available skills and knowledge and to decide how these should be employed. The work was undertaken with the Stone and Wood Carvers' Studio of English Heritage. It was critical to the success of the project that it involved craftsmen who had a sensitivity towards conservation and friable fabric and an understanding of the value of surfaces.

During the first week skills and techniques were tested on an undecorated section of plaster. The consolidation of the two sections of decorative plaster was undertaken during the second and third weeks. This work is described in this chapter.

The chapel at Cowdray House ruins as depicted shortly before the destructive fire of 1793. The altar covers a blocked window behind. To the left and right are modelled figures of Mary and Joseph. The whole scheme is in elaborate, modelled and cast lime plaster much of which was gilded. The plasterers, believed to be Italian, were craftsmen of a high order, and the work, even as it survives today, is of very high quality.
(From a photograph in the Cowdray House Ruins Museum)

CLEANING AND CONSOLIDATION OF PLASTER: CASE STUDY – PHASE 1

A similar view of the chapel taken in 1984 shows the devastating effect of the fire and nearly two hundred years of exposure to the weather. Remarkably, enough examples of detail remain to be able to recreate the original scheme almost in entirety. The brief for the conservation project was to retain everything possible and on the assumption that the roof would not be replaced.
(Photograph by courtesy of Corinne Bennett)

1985
During a two-week period in 1985, the RTAS team returned to Cowdray. Skills and techniques were developed further and three more areas of decorative and modelled plaster consolidated. These areas were higher up the walls and hence more friable. This work is described in Chapter 9.

1986
During three weeks in 1986, areas of particularly friable and structurally unsound plaster near the original ceiling line were consolidated (see Chapter 10).

8.2 THE SITE

The early eighteenth-century plaster in the chapel at Cowdray House ruins, the work of Italian plasterers, survives in fragmentary form only. The house was largely destroyed by fire in 1793, since when the plaster has been open to the weather. Even so, there is enough detail to establish the majority of the decorative scheme at least up to window head height. Supplementary information on the scheme is available from an eighteenth-century drawing and from photographs taken early this century. Recent photographs confirm that very considerable losses of plaster have occurred in the last sixty years, principally from upper levels. What survives is thus of enhanced value and considerable interest and its continued existence is of great historic and artistic importance.

8.3 CONDITION OF THE PLASTER

Without substantial consolidation there could be no question that much more of the Cowdray plaster would be lost in a very short period of time. During at least two separate periods remedial work in the form of cement:sand or cement:lime:sand mortar had been carried out. In many cases plaster losses over the last sixty years were recorded by empty mortar 'frames'. The installation of these mortar features had, in some cases, provided a support below and a certain amount of weather protection above. Unfortunately, in the majority of cases, their strength and relative impermeability had encouraged the deterioration of the plaster they were intended to assist. Frequently a fine crack had appeared at the mortar-plaster line and/or at the plaster-wall interface. Water access at these places had encouraged detachment of the undercoat plaster from the wall, the disintegration of some backing plaster, especially that apparently subjected to great heat during the fire, and the detachment of the finishing plaster from the backing plaster.

The finishing plaster was largely cracked and blistered, sometimes distorted and pitted. Some areas, especially near the ground, were damaged and disfigured with graffiti. Much of the surface was covered, sometimes thickly, with algae and crustose lichens; the lichen pads had tended to etch and soften the plaster on which they were growing. Small, hair-like roots of other plants had penetrated much of the undercoat and the top edges of many plaster panels supported moss and small lime-loving plants.

The modelled detail of the plaster was clearly of high quality and much of it appeared to have been painted or sized for gold leaf.

8.4 SCOPE OF WORK IN PHASE 1

- Plain plaster panel – north wall
- Decorative plaster panel – north wall
- Decorative plaster panel – south wall – 'first-aid' treatment
- All these areas were recorded photographically before work began

CLEANING AND CONSOLIDATION OF PLASTER: CASE STUDY – PHASE 1

One of the elaborate decorative panels before any remedial work was carried out. The main features of the conservation problem may be seen, including heavy organic growth, fragile top coat with detaching strap-work, crumbling undercoat and impermeable cement mortar patches and edge fillets.

8.5 INSPECTION AND ASSESSMENT

The inspection, which had to be preceded by some biocide treatment and preliminary cleaning, identified the following:

- Areas of plaster detached from the wall (hollow when tapped)
- Areas of finishing coat detached from the undercoat (hollow when tapped and springy to touch)
- Areas of detail unsupported and in imminent danger of loss
- Location of cracks between mortar fillets and plaster, or mortar fillets and wall
- Exposed edges

Based on the results of the inspection the following decisions were made:

MORTARS, PLASTERS AND RENDERS

- Where to introduce grout
- Where to introduce mortar fillets
- Where to patch
- Where to surface fill
- Where to apply limewater
- Where additional mechanical support was needed

8.6 GENERAL PRINCIPLES AND SEQUENCE OF REMEDIAL WORK

The following general sequence was established:

1. Treatment of plaster and adjacent masonry with biocide and removal of organic growth
2. Strengthening weak undercoats with limewater
3. First-stage mortar filleting of vulnerable edges
4. Flush-filling with mortar
5. Grouting detached interfaces within the plaster and between the plaster and the wall, providing mechanical supports of non-ferrous gauze and screws where needed
6. Removing all old cement-based fillets and patches
7. Filling holes with matching mortar (including grout holes)
8. Second-stage mortar filleting of remaining edges
9. Applying weather coat to undercoat plaster
10. Re-treatment of plaster and adjacent masonry with biocide

Items 3, 6, 7 and 8 took place in different sequences as conditions detailed. Item 2 continued intermittently during the course of the work

8.7 DETAILED DESCRIPTION OF REMEDIAL WORK ITEMS

Treatment with biocide
The very extensive colonization of all plaster surfaces, especially on the north-facing wall, with moulds, algae, lichen and mosses indicated that the role of a suitable biocide was very important. A biocide was used to establish conditions in which inspection and consolidation could take place and was recommended as a long-term maintenance treatment to maintain clean, sterile surfaces.

Cleaning and Consolidation of Plaster: Case Study – Phase 1

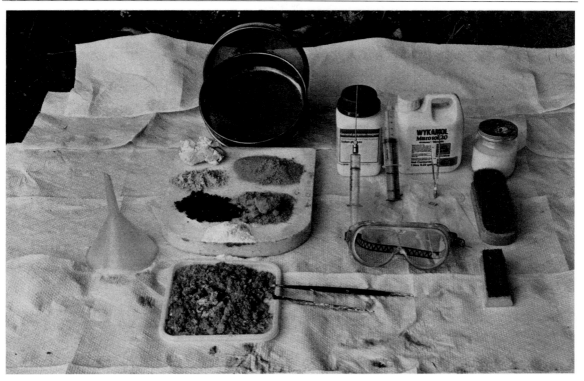

Typical selection of the conservators' tools and materials, left to right, includes grouting funnel, aggregate sieves, lime putty, stone dust, brick dust and sand, filleting mortar with dental and sculpting tools, hypodermic syringes, intrusion aids, biocide, safety glasses and bristle and phosphor bronze brushes.

The biocide selected was 'Murosol 20', a quaternary ammonium-based product with a proven, good record on Ancient Monuments sites. The concentrate was diluted with nineteen parts of water and was applied by hand sprays or by brush-flooding. The objective in each case was to ensure liberal treatment of the affected surfaces with minimum disturbance of the weak areas. In many cases it was possible to remove some lichens and mosses before treatment by gentle scraping with a spatula and brushing with small bristle or nylon brushes. Some areas were too fragile to be touched and cleaning had to be postponed until at least some consolidation had taken place. Small, compact brush heads of soft phosphor bronze wire were excellent at cleaning adjacent masonry but were too harsh for most plaster surfaces. Toothbrushes and stencil brushes were the most useful tools. Biocide application was carried out at least one day before cleaning. The dilution was always carried out immediately before use so that no diluted material was stored for more than a few hours; this was to avoid any risk of reduction in toxicity.

Very dry areas where the growth was markedly water-repellent were spray-wetted before treatment with water to make them more receptive to the biocide.

Removal of growth after treatment was time-consuming largely because of the small-scale tools with which the work was necessarily carried out; it was found to be absolutely essential to work 'with the detail' and not to use any system which scraped or brushed across modelled areas.

The cleaned surface was pitted with lichen scars and in some cases was stained, but all the detail read clearly and showed the full extent of the damage due to fire and weathering which needed attention.

Consolidation of undercoats with limewater

The soft, friable nature of much of the undercoat was a major difficulty and frequent applications of limewater were made throughout the early stages of treatment to effect some consolidation. Limewater was brought to the site in an airtight plastic container. (See Volume 1, Chapter 8, 'The cleaning and treatment of limestone by the "lime method"'.) All applications were made by hand-sprays. At first, the spraying tended to dislodge and transport some of the crumbly material, but a significant 'tightening' of the surface was achieved after the first day of application, allowing further useful treatment to take place without more losses. Reasonably accurate assessments of quantities and numbers of applications to given areas can be made but tend to be misleading. Limewater was applied from time to time by whoever was working in the vicinity of the treatment area, at a rate dictated by the take-up characteristics of the substrate. As an example, approximately 8.5 litres (2 gallons) were applied to an area roughly 0.75 metre ($\frac{3}{4}$ yard) square and achieved a significant consolidation effect.

Limewater consolidation was not applied exclusively to undercoat plaster but in general the fine finishing coat had a firm surface even when cracked and disrupted.

Mortar fillets: first stage

The function of mortar fillets around the edges of exposed plaster is to prevent or substantially inhibit the penetration of water into the body of the plaster where it tends to disintegrate the soft undercoat and to separate plaster coats from each other and the wall. The fillets also have an adhesive and supportive role. Mortar fillets were provided to all external edges of plaster and to the edges of lacunae within areas of plaster where it was decided not to flush-fill, or as first-aid or grouting dams. The mortar designed for fillets was colour-matched to the undercoat and was a relatively low-strength pozzolanic type based on lime putty and brick dust. (See 'Mortars' in the Appendix, this Chapter.)

When old fillets had been removed, or where none had existed, small amounts of plaster were removed by spatula and toothbrush to provide a reasonably sound edge. Where possible, the edges were undercut by spatula to provide more support and overcome the danger of water penetration at the fillet-plaster interface. The fillet splay angle from plaster face back to wall was, on average, 30° on upper surfaces and 45° on lower, showing the mortar edge ranging in thickness from 3–4 mm to 35 mm ($\frac{1}{8}$in–$1\frac{1}{2}$in).

Loose, powdery plaster was removed by blowing and gentle brushing and shallow lacunae were increased in depth to provide at least 6 mm ($\frac{1}{4}$ in) depth for filleting.

All areas, after preparation, were liberally sprayed with limewater. The excessive dryness of the old plaster and the wall made it imperative to persist with pre-wetting until the surfaces remained damp between spraying operations. Failure to achieve a truly damp substrate resulted in de-watering the fillet mortar and a pattern of shrinkage cracks. At least three separate applications of limewater were given to plaster edges with at least one-minute intervals. The water jet was also used in conjunction with a small spatula to remove further loose material in the fillet area. Any surfaces still black or green with algae were further treated with biocide and scraped clean with the edge of a spatula.

While the surface remained damp, fillet mortar was placed carefully around all exposed edges with a small spatula, modelling tool, pointing trowel or fingers, depending on the size of fillet. Only small amounts were placed at a time, working the mortar carefully under the edges of the plaster. In some cases 'sprung edges' of the plaster could be eased very gently back onto a full, soft bed of mortar, but in general filleting and filling took place to the plaster line as found, even where only slightly distorted. Much of the thin 3 mm ($\frac{1}{8}$ in) outer coat was quite brittle and would not tolerate pressure without fracturing. Large fillings were built up by packing in layers not exceeding 10 mm ($\frac{1}{2}$ in) at one time and were later re-wetted before the next layer was applied. In all cases rapid drying out must be avoided. In summer conditions the fillets and fillings may need to be covered with damp cotton wool packs to ensure slow drying.

The mortar was worked in towards the edge of the plaster with great care to ensure maximum contact by compaction without risking disruption. No mortar was spread onto the face of the plaster; a neat well-defined edge was always left and a clean line formed against the wall.

During application the spatula was regularly cleaned or dipped in water to reduce drag, encourage compaction and stimulate the formation of a small amount of lime laitence which could then be worked into the fillet-plaster interface. The technique of compacting the mortar was employed to reduce the risk of shrinkage and cracking.

The surface of the fillet was finished by compressing a damp sponge against it or by dragging the edge of a spatula across it; both techniques left a slightly texture surface intended to assist future evaporation of moisture from the fillets. Where shrinkage cracks had appeared in the new mortar by the following day the fillets were re-worked and compacted with no, or very little, additional water. Re-worked fillets rarely cracked again.

Flush-filling with mortar

Flush-filling of deteriorated undercoat was adopted where it was considered that more positive support than could be provided with mortar fillets was required or where surviving modelling could be better presented and more easily read against a flush background. In some situations, broken surfaces created confusing shadow lines which made it difficult to appreciate the design. The mortar for flush-filling was substantially the same as that for fillets, matching the colour of the original

A very large void requires building up with brass mesh and brass screws and several coats of lime and brick dust plaster. The mesh extends behind the bulging original plaster on the right and will later be grouted in to provide support. The plaster is supported and secured in its distorted position, as the top coat could not tolerate pushing back to its original line.

pink undercoat, but included some slightly larger brick aggregate. (See 'Mortars' in the Appendix.)

Areas for flush-filling were well-wetted by hand-spraying after sterilizing the surfaces with biocide. Most filling was built up in two coats each about 6 mm ($\frac{1}{4}$ in) thick, but shallow holes were filled in one. Thicker building-up was sometimes in three applications, each one being well ironed onto the substrate and scratch-keyed to receive the subsequent application. In one major hole which had been patched with a hard cement mortar it was necessary, after cutting away the cement, to fix 50 mm wide straps of brass gauze with brass screws into wall plugs bedded into the new undercoat plaster. The gauze passed into the voids behind the original plaster created by the disruption of the old surface. A bond between anchored gauze, new and old plaster was achieved by pouring grout or tamped mortar.

The final coat of flush-filling was finished by tamping with the ends of phosphor bronze brush bristles to leave a slightly raised texture. No 'restoration' in the form of modelling missing sections of decoration or moulding was carried out or intended. The exceptions were the completion of the line of decorative panels with new undercoat-matching plaster, in which case the perimeter was formed in a simple chamfer, and the modelled support mortar for pieces of fragile, unsupported detail such as the wing of the cherubim on the south wall.

Grouting

Probably the most critical operation in the exercise of cleaning and consolidation was the introduction of grout into the wall-plaster interface and the undercoat-finishing coat interface. The experience of the Moras (reference 3) and ICCROM research projects (references 1 and 2) was invaluable and the grout selection and grouting procedures followed, with minor modifications, recommendations based on Italian work. Grouting was carried out as follows:

1. Hollows were located by gentle tapping or by simple observations

2. Careful drilling was carried out with a small manual drill forming injection points 2 mm–3 mm (c.$\frac{1}{8}$ in) in diameter either directly into the plaster face (undercoat or finishing coat) or at an oblique angle to the wall, often through the existing cement mortar fillets or patches

3. Where detached plaster was free of the wall, loose fragments were removed by blowing out or by lifting out with tweezers

4. Using 15 cc glass syringes fitted with 1 mm blunt needles as the grouting tool, a wetting aid was introduced into open fractured and drilled injection points. The wetting aid was designed to facilitate the flushing of dry, dusty surfaces. A mixture of water and alcohol was used, coded 'FSA' (see Appendix: 'Flushing solution A')

5. Several syringes of water were introduced into the same points. During this operation, designed to wet up all the surfaces which would be in contact with the grout, the routes which the grout would follow were generally established. These routes were demonstrated either by the escape of water or by 'sweating' on the face of the plaster. Loose particles and water escaped through ports made at the lower edge of the void

6. At this stage, or at the commencement of the operation, clay plugs, mortar fillets and grout cups were placed in such a way that the grout could be contained where most needed. Enough exits were left to allow air to escape and to check on grout flows. Grouting was sometimes contained in a small patch a few centimetres square, at other times large cavities were extensive enough to be filled by pouring grout in; in some cases the grout travel was over a distance of up to two metres. Fragile areas liable to be detached by the pressure of grout were well supported with modelling clay or lime putty, the latter sometimes mixed with stone dust

7. A few minutes after washing out and just before the introduction of the grout into specific holes or open cracks predetermined by flushing, a further flushing solution was introduced. This was a mixture of water and

an acrylic emulsion coded 'FSB' (see Appendix: 'Flushing solution B'). The slight tackiness of this solution prepared the wall for the grout and was intended to prevent critical loss of water from the injected grout. Any escapes of 'FSB' were plugged with clay or lime and the surface cleaned at once with water

8   Grouting followed the flushing with 'FSB' immediately. The ingredients of the grouts are described in the Appendix (see Appendix 'Constituents and preparation of grout'). Ideally, the grout should be mixed mechanically. The procedure at Pompeii (reference 1) was to mix vigorously by hand followed by one minute of high velocity mixing in an electric blender. At Cowdray, all mixing was by hand and it was necessary to agitate the grout thoroughly before each draw-off and to keep the grout in the syringe moving while not injecting. To keep the piston of the syringe operating smoothly and to inhibit the trapping of fine solids between the piston and the glass body, clay slip or petroleum jelly was used as lubrication. Of the two pozzolanic additives tried, HTI powder remained in suspension longer than red brick dust, although both were ground to pass a 150 micron sieve

The rate at which grout was absorbed varied considerably. At times it was necessary to keep mixing fresh grout to feed one supply point; at other times one syringe of grout could take several minutes to be placed. The technique employed to contain the injected material in the plaster was to pass the needle of the syringe through a small sponge which could then be held, gently, against the plaster surface. A constant watch was kept at all times for any development of cracking or bulging which would indicate that the plaster was in distress. If such symptoms were apparent grouting was stopped immediately and additional support in the form of clay packs bridging the affected area between sound surfaces were applied. Where large areas need to be grouted it would be necessary to provide more positive support for the plaster face such as strutted plywood panels cushioned with soft felt. Deep pockets were grouted by pouring into clay cups or plastic funnels.

The grout had excellent penetration properties and cured within sixteen hours.

The large-voided panel on the north wall absorbed approximately thirty litres of grout poured and about sixteen litres placed by injection.

On the day following the grouting, previously hollow-sounding areas were frequently solidly bedded. Partially grouted zones, identified by further sounding, were opened with small drill holes and further grout was injected.

Runs and spillages of grout were cleaned off immediately using clean water or limewater spraying.

Removal of cement mortar fillets and patches
When grouting had been successfully completed and first stage filleting had been placed, it was possible to cut away the earlier cement mortar patches and fillets.

Partially cleaned detail of the panel previously illustrated. Having placed a lime and brick dust fillet around the edge of an area of plaster which has detached from its backing, the conservator is flushing and grouting using a hypodermic syringe working through an absorbent sponge cushion.

To guard against damaging the plaster the fillet was isolated by forming a small chase with a sharp chisel along the cement mortar-plaster line. Once isolation had been achieved the remainder could be safely cut away.

Secondary filleting
All edges exposed by the removal of old mortar were filleted in lime mortar as described above.

Plastering
All voids formed by the removal of cement mortar patches were plastered as described above.

Hole filling
All grout holes in the undercoat plaster were filled with the filleting mortar mix.

Application of weather coat to undercoat plaster

The filleting mortar was screened through a 150 micron sieve and gauged with a liquid consisting of skimmed milk and water in equal proportions. This was brush pressed to the undercoat plaster surfaces after consolidation with limewater and rubbed into the surface with a small coarse cloth pad. The pad was also used to compact the coat into the surface, following the technique of 'shelter-coating' as developed at Wells Cathedral and elsewhere. (Volume 1, Chapter 8, 'The cleaning and treatment of limestone by the "lime method"'.)

Application of mortar fillings and weather coat to finishing plaster

Small defects in the finishing plaster were filled with a mortar based on traditional Italian practice. This consisted of screened lime putty, gypsum and marble flour (see Appendix, 'Finishing coat mortar'). A weather coat of lime putty and gypsum was formed with skimmed milk and water and applied as described above. A smooth, semi-translucent, slightly polished surface was achieved by compaction and rubbing, providing a good contrast with the pink undercoat.

Second treatment with biocide

A further spray application of biocide was applied to protect the finished work from further colonization by destructive and disfiguring organic growth in the form of moulds, algae and lichen. This was particularly important where skim milk-gauged weather coats had been used.

8.8 SUMMARY AND RECOMMENDATIONS

The areas of plaster were successfully cleaned and consolidated and a method of working was satisfactorily established. The research exercise indicated that, contrary to initial assessments, almost all the surviving plaster, however fragile, could be held and consolidated. Detailed examination showed that the remaining work should proceed as soon as possible if further losses were not to be incurred.

Completed work on a section of one panel. The plaster shown is now solidly bedded on the mix of hydraulic lime, brick dust and acrylic emulsion, introduced as grout. The former cement mortar patching and edging has been cut away and, where needed, has been replaced with lime mortar filling and filleting. The whole area has been cleaned with biocide and given a thin protective lime weathering coat.

In addition to the areas described an inspection was made at high level of the window and blocked window reveals on which some of the finest plaster survives. These areas are plaster on lath. Although consideration was given to removing and refixing these sections on a new backing and was left as an option it was felt, finally, that effective treatment could be carried out in situ with less risk of loss.

The vulnerability of all the untouched plaster indicated that the work shown to be successful on the research exercise should be extended as soon and as extensively as possible. Priority was difficult to allocate but it was felt the modelled figures on the north and south walls should probably be amongst the first to receive treatment. *The work was found to be of such delicacy and of such a specialist nature that it was strongly recommended that it be completed by the same team which carried out the research.*

8.9 APPENDIX – SPECIFICATIONS

*All proportions are indicated in *parts by volume*, unless indicated otherwise.

Mortars

Undercoat and fillet mortar
(Mortar A)

Lime putty	1
Sand (1.18 mm)	$1\frac{1}{2}$
Bath dust (600 micron)	1
Brick dust (150 micron)	$\frac{1}{4}$
HTI (150 micron)	$\frac{1}{2}$

Finishing coat mortar
(Mortar B)

(For weathered, stained plaster)
As above but reducing brick dust by 50 per cent.

Adhesive mortar
(Mortar C)

(Applied as emulsion consistency)

Hydraulic lime (150 micron)	1
Sand (300 micron)	4
Brick dust (150 micron)	$\frac{1}{10}$
Acrylic emulsion	$\frac{1}{5}$

Adhesive no 2

Lime putty	9
Acrylic emulsion (10 per cent solution)	1

Finishing coat mortar
(*Mortar D*)

(For white plaster)
Lime putty	1
Marble flour (150 micron)	1
Gypsum	$\frac{1}{2}$

Grout

Hydraulic lime (300 micron)	1
HTI (300 micron)	$\frac{1}{4}$
Water	3
Acrylic emulsion	$\frac{1}{10}$
Sodium gluconate solution*	$\frac{1}{100}$ (increase by 100 per cent where extra mobility is required)

(*1 part sodium gluconate powder to 9 parts of water.)

Weather coats

Weather coat for undercoat
(*Mortar A*)

The fillet undercoat mortar is gauged with skimmed milk and poured through a 300 micron sieve.

Weather coat for stained finishing coat

Mortar B is gauged with skimmed milk and poured through a 300 micron sieve.

Weather coat for white finishing plaster

Lime putty	1
Gypsum	1
Gauged with skimmed milk:water	1:1

(*NB*: All milk-gauged material is treated on completion with the two applications of biocide to inhibit growth mould.)

Biocide

Quaternary ammonium biocide 'Murosol 20', 1 part of concentrate diluted in 19 parts of water.

Flushing solutions

FSA	Water	4
	Ethyl alcohol	1
FSB	Water	9
	Acrylic emulsion	1

Procedure for the preparation of brick dust

1. Crush large brick pieces to 5 mm ($\frac{1}{4}$ in) down with a club hammer on concrete
2. Put to dry on newspaper
3. Put dried material through 1 mm sieve ($\times 2$)
4. Grind 1 mm sievings with a pestle and mortar
5. Put 1 mm sievings through 500 micron sieve ($\times 2$). Dust used at this stage for pouring grade grout
6. Put 500 micron sievings onto marble slab for grinding under a marble pad
7. Put dust through the pestle and mortar for final grinding
8. Transfer dust to storage pot

Note: Freshly ground material was used for grout and mortar. If ready-ground dust is obtained in the future it should still be put through a pestle and mortar treatment or ground on a slab. The colour obtained from finely ground dust is noticeably darker than that obtained from the coarser grade.

Procedure for the preparation of grout

1. Put hydraulic lime through a 500 micron sieve
2. Mix hydraulic lime and brick dust from pot (see above) in the proportion 4HL:1 BD
3. Add dry powders slowly to water (see quantities above)
4. Add acrylic emulsion (see quantities above)
5. Add sodium gluconate solution (see quantities above)

REFERENCES

1. *Casa del Menandro, Pompeii*, Mission Report, ICCROM Research Training Unit 4, ICCROM, Rome, 1983.
2. Ferragai, Forti, Malliet, Mora, Teutonico and Torraca, *Injection Grouting of Mural Paintings and Mosaics*, ICCROM paper, Rome, 1984.
3. Mora, P, Mora, L and Phillipot, P, *Conservation of Wall Paintings*, ICCROM and Butterworths Scientific, London 1984.

9 CASE STUDY: CLEANING AND CONSOLIDATION OF THE CHAPEL PLASTER AT COWDRAY HOUSE RUINS – PHASE 2 (1985)

9.1 SCOPE OF WORK IN PHASE 2

The second phase of work at Cowdray House ruins during 1985 included plaster at higher levels of the walls, which had been made accessible by scaffolding. During this phase the following areas were treated as an extension to the research project commenced in 1984:

- Figure of the Virgin (?) and its surrounds, north wall
- Figure of St Joseph (?) and its surrounds, south wall
- A window reveal, south wall
- Linings to the blocked window above the altar, east wall
- Tiered drop, west wall
- Carbonation experiment on the lower levels of the altar wall, east wall.

With the exception of the absence of one apprentice, the research team was the same as the previous year, enabling full utilization and development of skills. Members worked singly and in pairs, depending on the size and complexity of the plaster in question. The techniques used were essentially those described in the previous chapter. Differences in approach dictated by the nature of the plaster or its deterioration, and any refinements or modifications to techniques are described in this section of the report.

9.2   TECHNIQUES USED

Before any work began the areas to be worked on were recorded photographically. This was followed by a treatment of biocide and the commencement of cleaning with toothbrushes and small spatulas. At the higher and more exposed levels at which the 1985 work was undertaken, it was noted that the plaster was considerably more weathered and hence extremely friable and fragile. In many areas even cleaning could not proceed without treatment with limewater until the plaster was touch-firm. Cleaning at times had to be coordinated into filleting and grouting work or on occasions postponed until after consolidation work was completed. The poor condition of many areas of plaster required careful grouting between the topcoat and undercoat, undercoat and masonry backing, and into fractures within the undercoat especially in the built areas of the modelled figures. In all situations the recommended sequence of flushings was carried out, even where the nature of the fracture meant it was not possible to flush out a void. It was considered better that a partial re-adhesion was achieved rather than none at all.

Grouting by syphon was used to fill larger voids. A hand-sized beaker was half filled with grout, held in one hand slightly above the hole to be grouted and gently agitated to prevent settling of solids. One end of a 6 mm ($\frac{1}{4}$ in) clear plastic tube was held at the bottom of the beaker, a syphon set up in the tube and hence the grout conducted through the other end of the tube into the void. The rate of flow could be finely controlled down to a slow trickle by raising or lowering the grout beaker. Whenever a leak was spotted the grouting end of the tube was raised above the height of the beaker and held in the same hand as the beaker, freeing the operative's other hand to plug the leak and quickly wash off any leaked grout.

A system of simultaneous filleting and grouting was used on sections of detached plaster which had vertical or downward facing edges which first had to be supported. The fillet was used as a dam for the grout, the sequence being conducted in 10–30 mm ($\frac{1}{2}$ in–$1\frac{1}{4}$ in) lifts, the next lift beginning when the previous grout had subsided and the previous fillet could be built onto. The fillets of deeper voids consolidated in this manner were built up in layers of 10 mm ($\frac{1}{2}$ in) maximum after the dam fillets and grout had taken on an initial set.

During 1985 the hydraulic set in the grout was achieved by HTI powder instead of brick dust. This produced a grout very similar in colour to the plaster undercoat which consequently did not produce disfiguring stains where it had leaked. The performance of the grout during placement was further improved by the sifting of dry components through a finer sieve (150 microns) than used previously. The problems of sedimentation and blockage of syringes were significantly reduced.

Extensive evidence of gold leaf and gold leaf size was found on the strapwork plaster of the south wall window reveal and the linings of the blacked window of the altar. The retention of this evidence required particularly careful cleaning, consolidation and repair as well as acceptance of a slightly soiled appearance in some areas. All such remaining evidence was spot fixed with an acrylic resin

(Paraloid B72) dissolved in toluene. (Gloves and masks were worn whenever this material was handled.)

During 1985 the procedure for removing cement fillets was refined. It was found that the bond between a fillet and original lime plaster could be best broken by gently working a hacksaw blade between the two. The blade was small and flexible enough to accommodate the varied profiles and the gentle sawing action provided a minimum of disturbance. Deep cement fillets could be isolated with this method before a hammer and chisel were used.

9.3   PHILOSOPHY OF APPROACH ON MODELLED FIGURES

The remains of the modelled figures on the north and south walls contained a significant proportion of modelled topcoat detail. There were also extensive areas of undercoat exhibiting different degrees of weathering and semblance of original form. It was necessary to consolidate the undercoat areas with lime watering and grouting. Lime mortar fillets were required to provide support to overhanging pieces and to prevent water penetration into and behind upward facing pieces. This work needed to be sensitively executed without detracting from the surrounding modelled work.

On the figure on the south wall the best support was sometimes provided by filling the line between two surviving points of a drapery or a moulding. Although this involved an element of restoration it was never carried to the extent of speculation or interpretation. It was significant that the quality of work produced resulted from work being carried out by a sculptor. On the north wall figure this approach was used to a lesser extent because less evidence of modelled surfaces remained. On both figures and their surrounds only original areas of modelling were treated with white weathercoat enabling these to be read against the repairs and surviving exposed undercoat areas which were treated with weathercoats coloured with Bath dust and staining sand to match.

9.4   SOUTH WALL WINDOW REVEAL

Work on the south wall window reveal involved the repair and consolidation of delicate areas of strapwork. Voids between the topcoat of plaster and the undercoat were usually small and would not accept much grout. As many as possible of these were grouted to provide a close network of 'spot fixings'. Wherever possible damages within the strapwork were used as grout holes.

As the strapwork was comprised of separately made pieces originally applied after the topcoat was completed, parts of it had detached and bowed away from the topcoat. Their under surfaces were filleted on a very small scale using a mortar of finely sieved lime putty and silver sand, grouted, and then the upper junctions were filleted to prevent water penetration. Segments of decoration such as petals which were seriously detached or had been held on by lichen, were

Mortars, Plasters and Renders

The modelled figure of Joseph nearing completion. A few good areas of detail only, survived the fire and subsequent weathering. Lacunae have been filled and details supported by grout and mortar fillets. The original surface alone has been protected by a fine white lime and skimmed milk based shelter coat. Below the figure the previous year's work is still being kept clean by the effects of the quaternary ammonium based biocide.

The decayed wood supporting the plaster lining has been cut away. Surviving oak lath has been treated with a wood hardener and scrim and has been bonded to the back of the old plaster to provide an additional key to the lath. New hardwood blocks screwed to the brickwork with brass screws are the anchor points for an armature of stainless steel mesh, being cut, here, to shape. An undercoat plaster was applied to the mesh and finished just below the line of the original face work. The completed detail is shown on the cover of this volume.

carefully removed, the contact surfaces cleaned and dampened and the pieces stuck back in position with a 'glue' of 9 parts lime putty and 1 part 10 per cent strength acrylic emulsion. Fine cracks in the strapwork were blown clean, dampened and filled with the finely sieved lime/silver sand mortar mentioned above or lime putty alone. To finish, the topcoat areas received a white weathercoat, the undercoat areas received a weathercoat coloured to match, and the entire panel was treated with biocide.

The restraint of time meant that the eastern side of the north wall window reveal received emergency consolidation work only. The plaster of this surface was substantially weathered and very friable. The panel was made touch-firm with approximately 50 treatments of limewater. It was then treated with the lime:casein shelter coat. This panel contains the best evidence of the methods of making and applying the chapel plaster strapwork and modelled decorations and it was recommended that it should receive attention during the 1986 programme.

MORTARS, PLASTERS AND RENDERS

Strapwork and cherub modelling on the head of the blocked window above the altar. The high quality of the original work can be seen in the fine detailing of the feathers. The dark areas are mostly a size base for gilding, although only very small areas of gold remain. This section is built up on riven oak lath and block, much of it attacked by wood beetles, and on iron nails. Much of the damage in this illustration emanates from the rusting of the nails.

9.5 LININGS TO ALTAR RECESS

The linings to the recess above the altar contained the most intact example of decorated strapwork and modelling in the whole chapel. The plaster of the curved head which was supported by laths and battens, had lost large segments. The aim of the 1985 work was to reinstate a firm support for the plaster and prevent water penetration to the rear of the panel by building up the missing areas in a manner appropriate to the form and design of the original work.

The back of the plaster head was first vacuumed, revealing the oak laths which were generally in good condition. Badly decayed wood of the other support battens was cut away and the remaining wood treated with wood hardener ('Xylamon'). The back of the plaster was then treated with PVA into which four layers of open-weave reinforcing hessian were laid, each layer of hessian running at 90° to the previous layer. New hardwood blocks were secured to the brickwork with plugs and brass screws in the areas where the plaster lacked support.

Missing sections of the soffit between the brickwork and the broken edge of the surviving plaster were formed in stainless steel expanded metal. On this, the plaster undercoat was built up in haired plaster, each layer being scratch keyed and not exceeding 10 mm ($\frac{1}{2}$ in) thickness. The finishing coat of lime:sand/Bath dust plaster was brought to the broken edge of original plaster, at the level of the top of the original undercoat, and finished with a wood float.

Flaking areas of gold size were again spot fixed with acrylic resin and a white, translucent sheltercoat applied to all original top coat areas. (The translucency was achieved by increasing the proportion of skim milk/water to lime putty.)

The vertical, side linings to the opening received emergency consolidation work comprising limewatering, some filleting and a weather coat, but required further work.

Cherub's head: altar recess soffit

The cherub's head at the top of the altar recess soffit had been badly split by the rusting of a wrought iron nail fixing and part of the features had been lost. When the new support plaster had been completed the nail was carefully excavated, withdrawn and replaced with a brass screw, the damages being made good with the undercoat plaster mix.

9.6 CARBONATION EXPERIMENT

Preliminary investigations were carried out into the possible acceleration of the carbonation of limewater and the curing of lime plaster/mortar by creating a carbon dioxide rich atmosphere. This was achieved by erecting a simple polyethylene sheeted frame a few centimetres from the plaster face. Seals were formed by lapping and taping sheets. Carbon dioxide from cylinders was introduced at the base of the tent so that old limewatered plaster and new lime plaster were exposed to the CO_2. Early results were promising and further investigations are planned.

10 Case Study: Cleaning and Consolidation of the Chapel Plaster at Cowdray House Ruins – Phase 3 (1986)

10.1 BACKGROUND TO THE 1986 WORK

The third and final phase of work on the decorative plaster of the chapel at Cowdray House ruins was undertaken to conserve and secure the remaining areas of plaster in the chapel. As these areas were generally located on the higher and more weathered parts of the walls, they were in the most friable and unsound condition of all the sections of remaining plaster. The experience of the previous two years was invaluable and enabled the retention of a surprising amount of highly delicate and significant plaster.

The time at Cowdray in 1986 also enabled the work of 1985 and 1984 to be evaluated.

10.2 SCOPE OF WORK

The scope of work included the following zones of plaster:

The cast frieze on the western wall
This frieze is composed of cast plaster sections approximately 225 mm × 300 mm (9 in × 12 in) which were planted into a lime mortar base to form a continuous band with a repetitive pattern (see photograph). Similar friezework existed elsewhere in the house but the chapel section is by far the longest and most intact section to remain.

The curved head and reveal of the window on the south-east splay wall
The curved window head contained some of the best details of the remaining plaster. Its protected position also meant that extensive areas of paint and gilding size, blistered in the fire, also remained. The left-hand reveal to the window was made up of plaster on lath and studs which had been used to build the surface to the required profile. This area was very friable and structurally insecure as the studs were no longer fixed to the wall. An immense amount of dirt had accumulated behind the panel and was helping to wedge it off the wall.

The plaster with a panelled design on the north wall
This area of plaster approximately $2.7\,m^2$ (9 ft × 9 ft) and originally accessible by a gallery contained a design of run mouldings which imitated a panelled dado with hinged shutters above.

Between 60–70 per cent of the body of the plaster was detached from the wall. There were many separations between the various undercoats and top coats. Many of the applied mouldings were precariously attached and the only remaining section of a heavy moulding, run onto a brick corbel, was loose.

The piece of large frame between the panelled area and figure (south wall)

10.3 WORK PROCEDURES

The work procedures adopted in 1986 were based on and developed from those of previous years.

All areas were first treated with a quaternary ammonium biocide. Lichen, dead roots and other surface dirt were removed with the aid of small spatulas, toothbrushes and small phosphor bronze brushes. Some areas could not be cleaned until they had been consolidated. Limewater consolidation was carried out on all friable surfaces, even before cleaning was completed if necessary.

Detailed descriptions of the works follow. Just before the site was vacated, all areas received a further biocide treatment.

10.4 DESCRIPTIONS OF THE AREAS OF WORK

The cast frieze (west wall)
The cast plaster segments of the west wall frieze were in very good condition. The cement-based fillet along the top of the frieze was cracked and fractured and had allowed a lot of dirt to wash in behind the sand:lime render which formed the base for the cast pieces. This was also an area of damaging frost activity.

All the cement filleting was removed from around the frieze. The void along the top was blown clean and grouted. Root activity of a previous ivy covering had dislodged several small pieces which were re-adhered with the 'gluing' mortar developed in 1985 (1 part 10 per cent strength sodium gluconate solution to 9

Mortars, Plasters and Renders

The cast frieze under the missing balcony. Repetitive, cast detail is unusual at Cowdray but this, too, is of considerable quality. The frieze acts, now, as a catchment for a large area of wall covered with organic growth, and discolours quickly. When work commenced it was covered with thick mould and lichen growths. To prevent rapid recurrence, as much wall face as possible above the plaster was sterilized.

parts lime putty). The plain plaster areas surrounding the frieze were consolidated as well.

Replacement and new fillets were formed in a mortar of lime putty, sand, Bath stone dust and brick dust gauged with HTI, as used in previous years, coloured to match the undercoat plaster. The treatment was completed with the application of a white weather coat.

The window head and reveal (north-east wall)

The support structure of the window head was not visible but appeared to be adequately supporting the curved plaster section. The several cracks in the plaster did not appear active and may have related to times when the top surface of this area was exposed to the weather. Water penetration also appeared to have been the cause of the exposure of several areas of undercoat due to the loss of areas of top coat and modelling. The blistered paint on about 50 per cent of the remaining top coat areas was in a range of thicknesses whose condition ranged from sound to too friable to touch with a sable brush.

The remaining details, condition, structural soundness and relative cleanliness of the window head meant that a minimal amount of work was done to this area.

All areas of paint which did not fracture under the gentle touch of a sable brush were consolidated with Paraloid B72. Areas of friable undercoat could then be treated with limewater and have their cracks cleaned out, slightly undercut, pre-wetted and filled with a colour-matched mortar. The undercoat areas then received a weather coat treatment.

The window reveal contained no decorative detail apart from at its junction with the curved head; its plaster comprised a 1.5 m × 0.75 m (5 ft × 2 ft 6 in) area of friable undercoat. The work undertaken on it was primarily aimed at structural integration of the plaster and support structure, the whole area was barely supported and could move freely.

First of all it was necessary to remove all pieces of decayed timber and other finer dirt from behind. This was done with long pieces of wire and required the removal of a cement edge fillet and the formation of two small holes through the undercoat. Only then was it possible to fix the main supports to the wall, and to fix several laths to these supports. Wood hardener was then applied to all accessible timbers as these were heavily weathered.

The junction between the plaster and the wall on the window side was built up in mortar. The other side of the plaster required a far more complicated procedure as here the ends of many of the laths had rotted out. Expanded stainless steel mesh was inserted into these areas, tucked in under the edge of the original plaster and fixed to remaining laths. The missing area of plaster could then receive a mortar fillet which linked the support structure to the plaster once again.

Once structural soundness was achieved, cracks in the plaster were raked out, undercut and filled, the edges were filletted and the access holes were filled, all with a matching mortar. The reveal plaster then received in excess of thirty coats of limewater followed by a colour-matched weather coat.

The panelled area (north wall)

Consolidation of the panelled area of plaster required more grouting than any area previously undertaken. Work began at the bottom of the area and required the labour of three people for two weeks for completion. After biocide treatment and initial cleaning, limewatering, filleting and grouting began. The large cement-based fillet on the perimeter of the area was removed in short sections and only when it was certain that the area of plaster it supported had been grouted.

The panel was formed by 2–3 layers of undercoat and a skimmed top coat onto which the run mouldings of the framing were planted. The areas outside the panels formed were built up further with undercoat and a further skimmed top coat. This provided plenty of scope for interface failure and extensive areas requiring very fine filleting and grouting between the various layers. The mortar used for filleting in such areas contained well-graded aggregate which was significantly finer than that in the mortar used for the larger fillets (less than 600 microns).

During the last phase of the plaster exercise at Cowdray the panelling to the gallery was conserved. As before, grouting, limewater consolidations and mortar filleting formed the bulk of the work, finishing off with shelter coated undercoat and top coat in different colours to match the original. The architectural detailing is quite sophisticated, down to the modelling of false butterfly hinges and nail heads.

Limewatering was carried out during the entire treatment.

Because of the large size of the voids, support was first provided with timber strips fixed through the plaster to the wall behind. Many parts of the voids had filled up with dirt and friable plaster. Special effort was made to flush these as clean as possible to ensure maximum adhesion.

Several sections of run moulding had detached from the body of the plaster and were in a fractured and distorted form. It would not have been possible to refix these in their original positions without the loss of original fabric. Such pieces, therefore, were consolidated in their distorted form by filleting and grouting.

The treatment of the panelled area finished with the application of a colour coordinated weathercoat. The original top coats of plain areas and moulded sections were treated with a white coat to match their original surface colour. Parts of the original cream-coloured undercoat had turned pink in the late eighteenth-century fire. This was matched with a pink weather coat with a cream coat on undercoat areas not affected by the heat.

The large frame
The frame comprised a 2400 mm (8 ft) section of plaster built up on brickwork with overall profile dimensions of 225 mm × 150 mm (9 in × 6 in), most of which was no longer attached to the wall. This was achieved with a combination of filleting and grouting. The surface was then consolidated with limewater, fillets and a weather coat.

10.5 ASSESSMENT OF THE COWDRAY PROJECT

The pioneering Cowdray plaster project of RTAS has generally been very successful. About 99 per cent of the original and highly friable plaster of the chapel has been conserved in a technically, philosophically and aesthetically acceptable manner which will assist it to survive in its external environment. It is now possible to receive a very good impression of the original decoration of the Cowdray Chapel. It is also possible to record all surviving detail accurately.

It was important that the team who undertook the work initially had a good understanding of the value of surface finishes and details and the appropriate delicate manual skills. This understanding and specialist craft ability was critical to the assessment of the situation at Cowdray and the correct application of experience and theory from elsewhere. RTAS was fortunate to have a consistent core of people who were on the job each of the years and provided the basis for learnt skills to be developed and refined.

At the beginning of the project it was decided that every attempt should be made to keep every piece of plaster, no matter how bad it looked. This provided a challenge to all involved and motivated the refinement of skills to the high level that was achieved. It also proved clearly that no assumptions on time, materials, skill levels and costs for this sort of work should be made before a thorough, practical feasibility study has been carried out.

11 CASE STUDY: REMEDIAL WORK TO SECURE GRAFFITI ON PLASTER

Condition survey carried out at Richmond Castle cell block by HBMCE Research and Technical Advisory Service on 12–13 March 1985

11.1 BRIEF

A two-storey cell block lies just within the curtain wall of Richmond Castle, Yorkshire. During the First World War (1914–18) this military prison was used to confine 'prisoners of conscience', those who for religious, political or other reasons refused military service. Between the two wars and during the Second World War (1939–45) the cells reverted to regular army prison use and for storage.

The plastered and painted walls of many of the cells bear interesting examples of pencil graffiti left by prisoners, especially of the 1914–18 period. Whilst many of these are in remarkably sound and clear condition others are lost or partially lost due to the dampness of the walls and the vulnerability of the distemper which forms the ground for the graffiti.

The Research and Technical Advisory Service were asked to carry out an investigation into the environmental and surface problems. This involved establishing moisture contents of the building fabric and enclosures and carrying out preliminary investigations into consolidation and cleaning.

11.2 PROCEDURE

1 Moisture content in the walls and roof/floor slabs was measured using a 'Speedy' carbide unit. Samples were drilled out and moisture contents measured on the spot. The location of these samples and the moisture contents are shown on Figure 11.1.

2 Moisture humidity of the cells and corridors was measured during the afternoon of 12 March and during the morning of 13 March. Duplicate readings

REMEDIAL WORK TO SECURE GRAFFITI ON PLASTER – CASE STUDY

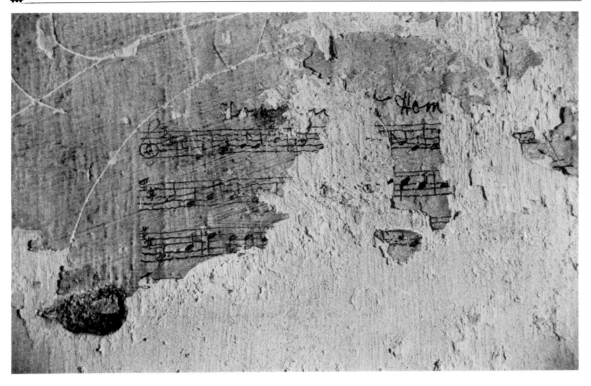

Sometime between 1914 and 1918 a prisoner of conscience at Richmond pencilled the score of 'Home, Sweet Home' on the distempered wall of his cell. This and many other examples of graffiti, some of considerable social and political interest are to be conserved by a combination of environmental control and fragile flake-laying using a mixture of alkoxysilane and acrylic resin.

were taken using a whirling hygrometer and a Vaisala unit, measuring relative humidity and air temperature.

3 Small-scale tests with consolidants were carried out as described below to fix flakes of distemper back to the plaster.

4 Trial cleaning was carried out with a quaternary ammonium biocide, a neutral pH soap and 'molecular trap' rubber.

5 On completion, a plastic catchment sheet was taped out on the floor of cell 6 to establish the current amount and type of loss occurring under the most severely deteriorated area.

11.3 DISCUSSION AND RECOMMENDATIONS

The survey established that the building was now 'dry' due to remedial work to the roof with the positive exception of those areas adjacent to the east wall and

Mortars, Plasters and Renders

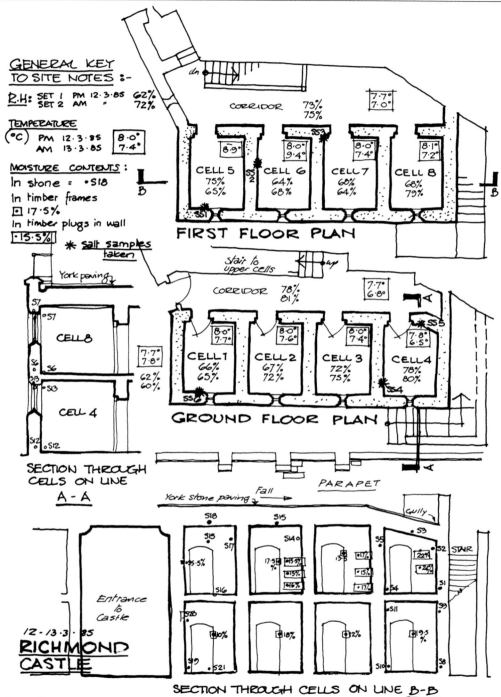

Figure 11.1 Site notes relating to plaster/paint consolidation

the south-east corner. The extreme dampness of this area was the result of the staircase which had been acting as a drainage route for the roof water. Dispersed water from the roof which, in the past, percolated or streamed through the butt-jointed sandstone roof slabs and which was responsible for all major losses of graffiti, was now being satisfactorily excluded.

Unfortunately, a residual problem was left behind from past saturations. Plaster in the old saturation zones was friable and in some cases had been lost; where it remained intact the thin layer of distemper on which the pencilled graffiti lay tended to detach itself in thin, papery scales. In these zones there were residual salts which were probably hygroscopic within the likely range of temperatures in the building. Salts were removed from key areas for identification and measurement of hygroscopicity. The chief problem arose in spring and summer when the wall temperature was markedly lower than that of the surrounding air; even when the building was empty moisture condensing on the wall surfaces was likely to be destructive to fragile flakes, and large numbers of summer visitors would certainly exacerbate the problem. Increases in temperature and relative humidity were quite substantial during the survey when three people were working in the confined space of a single cell.

Other recorded damage to plaster, modern and historic, was less complex. Graffiti of the 1939 period included scratched designs which had the effect of flaking off the old distemper; wear and tear of what was considered a general utility building had taken its toll. Cell 1 was much disfigured by paint. In this situation it was likely that careful work with water-rinsable methylene chloride, scalpel flaking or even a small air abrasive unit would be successful in revealing currently obscured graffiti, although these techniques were not applied during the current investigation.

The scale of the main problem, in the old saturation zone areas, was not large but significant losses of interesting material were likely. The critical zones were:

- Cell 8 east wall and north wall
- Cell 6 east wall
- Cell 4 west wall

These were the areas likely to suffer most from humidity and soluble salts and it was recommended that they be treated as described under 'Consolidation'.

Elsewhere and generally, the current environment was good and there appeared no need for substantial change. A thermohygrograph for installation in cell 6, the most critical zone, was set up to provide a record for the foreseeable future, so that any future problems could be related to seasonal changes and visitor influxes. Only on the basis of long-term observation could any modifications to the environment be made in the form of additional ventilation or de-humidification.

General recommendations arising from the survey may be summarized as follows:

- Complete the survey of all surviving graffiti photographically and by transcribing all the text
- Let a contract to specialist conservator to carry out consolidation of

critical zones as suggested below and to commence paint removal in cell 1

- Install a thermohygrograph in cell 6
- Carry out proposed works to external staircase
- Proceed with making good, repairing plaster, cleaning and treating iron and wood (cell doors).

The recommendation was made that it would be unwise for visitors ever to have access to cells where graffiti survives, even after consolidation, because the situation would always entail unacceptable risk.

11.4 CONSOLIDATION

The moisture content survey indicated that there was no reason why consolidation treatment of fragile, flaking areas should not proceed with the exception of the east wall of cells 4 and 8. As soluble salts are only likely to be activated by high relative humidity, it was important that consolidation should be carried out before any major influx of visitors.

The following consolidation treatments were tried on small areas and all had potential as 'flake-laying' systems:

1 Acrylic emulsion (Plextol) in water (10 per cent)

2 Acrylic resin (Paraloid) in toluene (5 per cent)

3 Acrylic resin (Paraloid) in acetone: IMS 50:50 (5 per cent)

4 Acrylic resin in silane (Raccanello)

The most effective was thought likely to be (4) with (2) or (3) on larger areas. A very slight initial yellowing was observed after resin applications but tended to fade in a short time.

The consolidant was applied from a beaker with a 5 mm flat bristle brush. The loaded brush was first held against the wall behind the flake. Because the plaster was dry it took up the consolidant very readily. The edge of the distemper flake was then touched lightly with the tip of the brush and took up consolidant. After a few minutes of this feeding procedure the flake was laid back onto the plaster. However, in some cases an accumulation of salts and powdered plaster wedged behind the flake and prevented contact; pressure would have broken the flake, which was relatively brittle. Where graffiti was at risk, therefore, fine tweezers and small paint brushes dipped in solvent were used to clear behind the flakes. This was extremely delicate work and in some cases was carried out under illuminated magnification.

Brush feeding or gentle spraying with acrylic resin in acetone/IMS was used on exposed edges of distemper where there was no graffiti.

Disrupted areas where there was no graffiti were carefully cut away with a sharp scalpel, laying the blade close to the surface as though making a paint scrape. These excisions needed to be only 1–2 mm deep and in releasing loose, powdery, flaking material relieved the tension on more important adjacent areas.

It was recommended that no overall treatment should be applied to the wall surfaces.

11.5 CLEANING

Much of the distempered surface was grimy and dusty. After first-aid fixing of flakes careful cleaning proceeded as follows:

- Stiff bristle brushes were used on the sandstone soffits of roof and floor slabs and window reveals to remove efflorescent salts and loose scales
- Soft paint brushes and hand vacuum brushes were used very gently to remove all loosely adherent dust from the plastered surfaces
- Neutral pH soap in water was used with small cotton swabs to clean the surface of sound distemper or plaster
- 'Molecular trap' rubber was used on small areas to pull dirt from difficult areas by very light pressure.

Making good

Plain plastered surfaces, principally in the upper corridor area and stair well were to be made good by plastering in damaged areas with lime:plastering sand 1:3. Cracks elsewhere were to be filled with the same material after flushing out with clean water and hand sprays. Where redecoration was to be considered it was recommended that a cement–size distemper, compatible with that used in 1914 was used.